100+

PRINCIPLES OF GENETICS

100+
PRINCIPLES·OF
GENETICS

ANTHONY J. F. GRIFFITHS

JOAN McPHERSON
The University of British Columbia

W. H. FREEMAN AND COMPANY
New York

Library of Congress Cataloging-in-Publication Data
100+ principles of genetics/
  compiled by Anthony J. F. Griffiths and Joan McPherson.
      p.   cm.
   Includes index.
   IBSN 0-7167-2016-7
   1. Genetics—Handbooks, manuals, etc.   I. Griffiths, Anthony
J. F.   II. McPherson, Joan.   III. Title: 100+ principles of genetics.
IV. Title: One hundred plus principles of genetics.
QH430.A15   1989
575.1—dc19   88-39075
                                                                       CIP

Printed in the United States of America

1 2 3 4 5 6 7 8 9 0   VB   7 6 5 4 3 2 1 0 8 9

CONTENTS

INTRODUCTION

100+ Principles of Genetics provides an easily accessible overview of genetics. We believe it will be especially useful as a supplement to textbooks in fundamental genetics. It could also serve to update or review a basic understanding of genetics.

Genetics today is an immense and complex subject, and in studying it it is often difficult to "see the forest for the trees." Nevertheless, there are threads that run through many parts of the subject. There are not many of these threads, and they are often quite simple. They nevertheless constitute the subject of this book. Some of these threads deserve the lofty status of a principle, such as the use of recombination in mapping. Others, though less imposing are nevertheless very important in certain areas, such as mapping human chromosomes using chromosome retention in human/mouse cell hybrids.

We have not been too fussy about what is or is not a true principle. We have, instead considered these more as important statements about genetics than as precepts chiseled into stone. Doubtless many will disagree with the way these statements have been organized, but such challenge and rethinking is useful in itself, and potentially a very efficient way of focusing on the essence of the subject.

The principles have been written with a view to being read in sequence. However, each could stand by itself and many cross references have been included to assist entry at any point. These appear as bold numbers within the text.

The general topic groupings are: genes and inheritance, mutation, gene structure and function, recombinant DNA technology, organelle genes, quantitative genetics, and population genetics. These groupings are not supposed to reflect any rigid compartments within the subject of genetics (these are more and more disappearing) but rather form a convenient progression for explanatory purposes. Furthermore, the groupings have not been adhered to tightly when convenient explanation demands a different approach.

We assume a familiarity with the basic terminology of genetics and cell biology, but many terms are explained in the process of expounding the principles.

We extend our thanks to the following reviewers: Edward M. Berger, Dartmouth College; Jerry Feldman, University of California, Santa Cruz; Richard L. Garber, University of Washington, Frederick Jay Gottlieb, University of Pittsburgh; David E. McMillin, Georgia State University; David Sheppard, University of Delaware; and Philip J. Snider, University of Houston.

Anthony J. F. Griffiths
Joan McPherson

PRINCIPLES

1

The raw material for genetics is individual variation.

There are two basic types of variation.

DISCONTINUOUS VARIATION

The existence in a population of two or more distinct forms, or **phenotypes,** of a given character.

When there are two or more *common* distinct forms, there is said to be **polymorphism.**

When one form is common and other very rare (e.g., less than 1 percent) the rare forms are called **mutants,** and the common form is called **wild type.**   **5 107 108**

CONTINUOUS VARIATION

Many characters in nature vary across a wide range, showing no breaks or discontinuities.   **104 105 106**

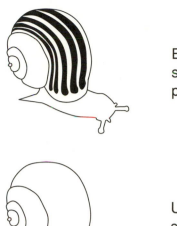

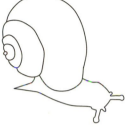

Banded
shell
phenotype

Unbanded
shell
phenotype

Discontinuous variation in natural populations of the snail *Cepaea nemoralis*. These and other banded phenotypes are common; hence, it is an example of polymorphism.

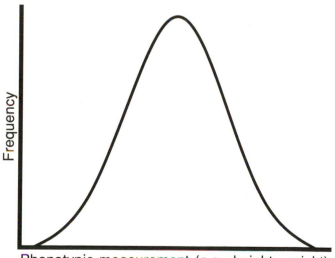

Continuous variation in one phenotype represented graphically.

2

Genetics is the study of genes through their variation.

A **gene** is a fundamental unit of hereditary information.

Humans have an estimated number of about 50,000 different kinds of genes in each cell, fruitflies about 5,000, and *Escherichia* (*E. coli*) about 1,500.

Each gene can take several different variant forms called **alleles.**

Allelic gene variation is a major cause of variation in a population, whether discontinuous **1 107 108** or continuous **1 106** .

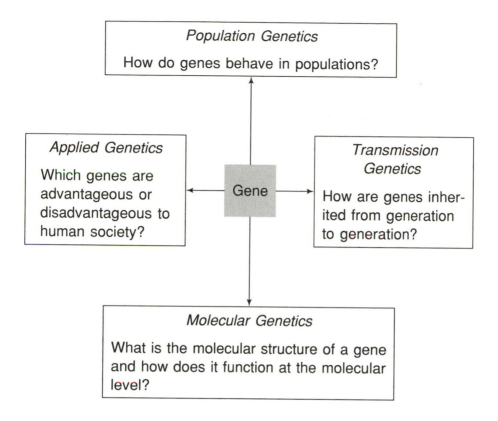

In each of these areas of genetics, gene variants play a crucial role in the modus operandi. The use of variants characterizes the genetic approach to biology.   **5 45 47**

3

Each cell of an organism contains at least one basic set of genetic information for that organism. That set is called a genome.

The genetic information is organized into units called **chromosomes** (**5**) and these are different in form and number in prokaryotes, viruses, and eukaryotes.

PROKARYOTE	VIRUS
Genome	Genome
circular chromosome	circular chromosome
O ×1 to several	O ×1 to several
	or linear chromosome
(e.g., bacterium)	(e.g., bacteriophage)

NOTE: Bacteria and eukaryotes can have additional extra-genomic independently replicating fragments called **plasmids,** which can be circular or linear **29 30** .

EUKARYOTE

	haploid $(n)^*$ cell	Diploid $(2n)$ cell
Nuclear chromosome set	×1	×2
	×1	×2
	×1	×2
	×1	×2
Organelle chromosome	× many	× many
	(e.g., yeast, human gamete)	(e.g., human body [somatic] cell)

genome

*n represents the haploid nuclear chromosome number.

4

Each gene has a characteristic chromosomal place (locus) in the genome.

In many cases, the location of a gene affects its operation **80** . A knowledge of gene location is essential for routine genetic manipulations **64** and for isolating genes **96**.

Example Bacterium

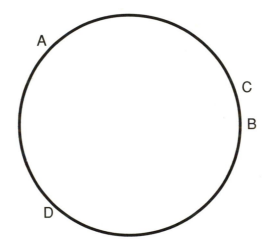

Genes *A, B, C, D* and their various alleles are always in same locations in this species.

Example Nuclear chromosomes of a *2n* (diploid) eukaryotic cell

NOTE: One chromosome set is derived from one parent, and the other chromosome set from the other parent. Hence the chromosomes are in **homologous pairs.**

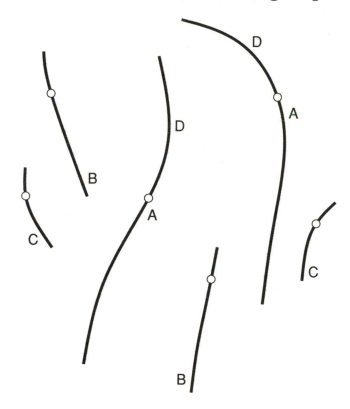

In such a diploid, each locus may bear identical alleles (**homozygous,** e.g., *AA* or *aa*) or different alleles (**heterozygous,** e.g., *Aa*).

5

In eukaryotic organisms, genes can be identified by observing Mendelian ratios of phenotypes in standard crosses involving discontinuous variants.

This principle defines a standard modus operandi for discovering the existence of a gene. We use an example actually studied by Mendel.

Logic

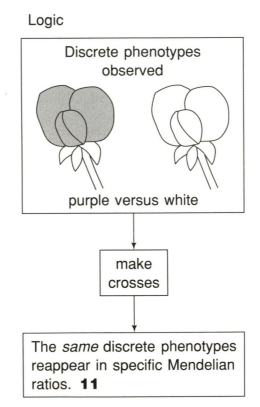

In prokaryotes and viruses, the logic is similar except that the ratios are not Mendelian. **28 31**

For example, in one of Mendel's crosses in peas, the results were as follows:

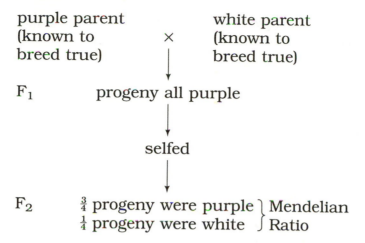

purple parent (known to breed true) × white parent (known to breed true)

F_1 progeny all purple

selfed

F_2 $\frac{3}{4}$ progeny were purple $\Big\rbrace$ Mendelian
          $\frac{1}{4}$ progeny were white $\Big.$ Ratio

Interpretation at the gene level is as follows:

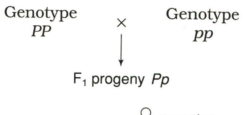

Genotype *PP* × Genotype *pp*

F_1 progeny *Pp*

♀ gametes

♂ gametes	$\frac{1}{2}$ *P*	$\frac{1}{2}$ *p*
$\frac{1}{2}$ *P*	$\frac{1}{4}$ *PP*	$\frac{1}{4}$ *Pp*
$\frac{1}{2}$ *p*	$\frac{1}{4}$ *Pp*	$\frac{1}{4}$ *pp*

F_2 Totals $\frac{3}{4}$ *P–* (purple)
$\frac{1}{4}$ *pp* (white)

P and *p* are said to be alleles. Such results are effectively a test for allelism.

There are also molecular ways of identifying genes **16** .

6

The phenotype of the heterozygote defines the dominance relationships of genes.

A knowledge of the way the alleles of a gene interact in the heterozygous condition can provide clues about their function **40** .

A heterozygote A_1A_2 potentially can show a variety of phenotypic expressions in relation to the homozygotes A_1A_1 and A_2A_2.

Example Complete dominance of A_1 over A_2

$$\left.\begin{array}{l} A_1A_1 \\ A_1A_2 \end{array}\right\} \text{ purple}$$
$$A_2A_2 \quad \text{white}$$

Example Incomplete dominance of A_1

$$A_1A_1 \quad - \quad \text{red}$$
$$A_1A_2 \quad - \quad \text{light red}$$
$$A_2A_2 \quad - \quad \text{white}$$

Example Codominance of blood group alleles

$$L^M L^M \quad - \quad \text{group M}$$
$$L^M L^N \quad - \quad \text{group MN}$$
$$L^N L^N \quad - \quad \text{group N}$$

IN GENERAL

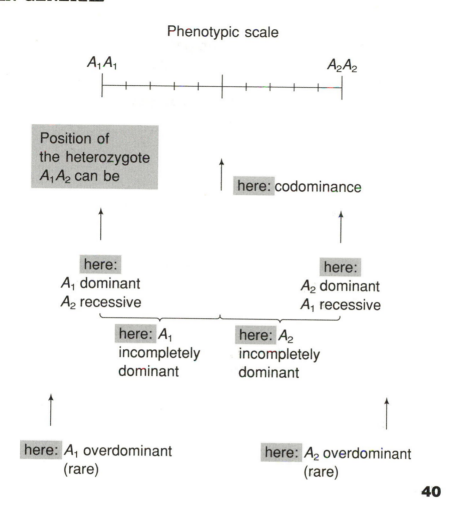

Phenotypic scale

A_1A_1 A_2A_2

Position of
the heterozygote
A_1A_2 can be

here: codominance

here:
A_1 dominant
A_2 recessive

here:
A_2 dominant
A_1 recessive

here: A_1
incompletely
dominant

here: A_2
incompletely
dominant

here: A_1 overdominant
(rare)

here: A_2 overdominant
(rare)

40

7

In eukaryotes, mitosis partitions the chromosomes equally into daughter cells during somatic cell division, and meiosis halves the number of chromosomes during production of sex cells.

Mitosis operates during the increase of cell numbers (e.g., development in multicellular organisms, or growth of a bacterial culture). Meiosis operates during the production of sex cells (e.g., formation of gametes, or of sexual spores such as ascospores).

MITOSIS
(in a diploid cell)

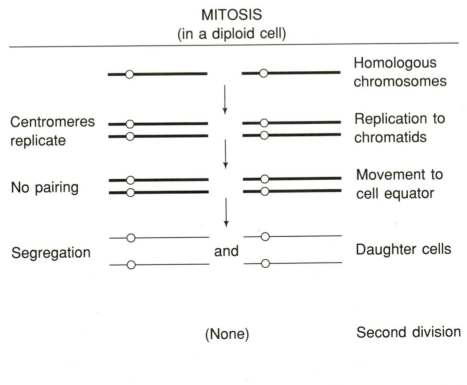

Centromeres replicate

No pairing

Segregation

and

(None)

Homologous chromosomes

Replication to chromatids

Movement to cell equator

Daughter cells

Second division

MEIOSIS **10**
(in a diploid cell)

Homologous
chromosomes

Replication to
chromatids

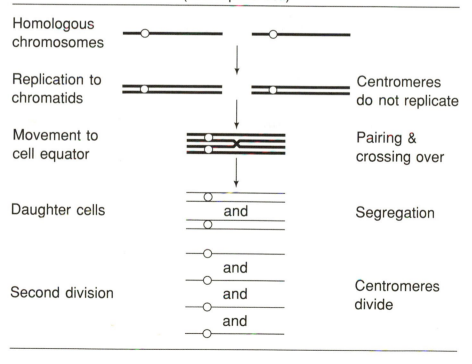

Centromeres
do not replicate

Movement to
cell equator

Pairing &
crossing over

Daughter cells   and   Segregation

Second division   and   Centromeres
divide

and

and

8

At meiosis, the members of a gene pair segregate *equally* into the meiotic products.

Meiosis can occur only in a diploid cell, where there are two complete sets of all genes, and hence each gene is a member of a pair.

Example   The principle operates on homozygous or heterozygous gene pairs, but can only be detected practically in heterozygotes.

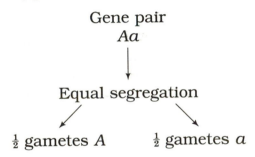

Gene pair
Aa

Equal segregation

$\frac{1}{2}$ gametes A $\frac{1}{2}$ gametes a

This principle is sometimes known as **Mendel's first law.**
5 9 10 11

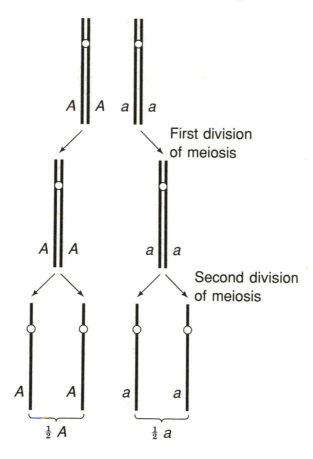

First division
of meiosis

Second division
of meiosis

$\frac{1}{2}$ A $\frac{1}{2}$ a

9

At meiosis, the segregation of one gene pair is *independent* of the segregation of gene pairs on other chromosome pairs.

This principle is sometimes called **Mendel's second law.**

Example Gamete production in a double heterozygote *AaBb* where the gene pairs are on separate chromosome pairs

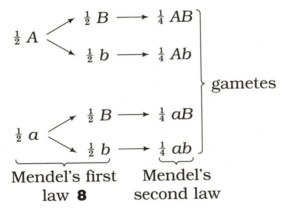

Mendel's first law **8** Mendel's second law

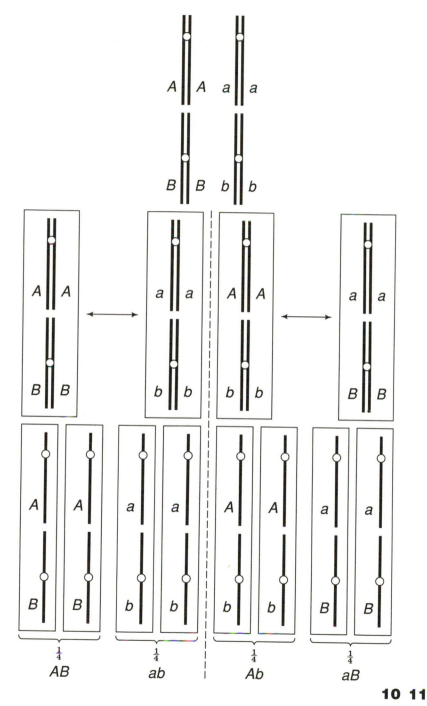

10

Mendel's laws (8 9) apply to any eukaryote in which a regular meiosis occurs.

In eukaryotes there are three basic types of life cycle:

- Haploid
- Diploid
- Alternating haploid/diploid

HAPLOID

For example, *fungi* (such as yeast and molds)

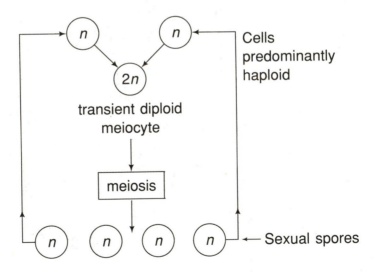

DIPLOID

For example, *most animals*

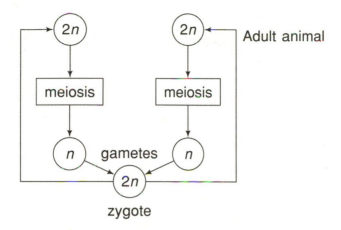

zygote

ALTERNATING HAPLOID/DIPLOID

For example, *plants*

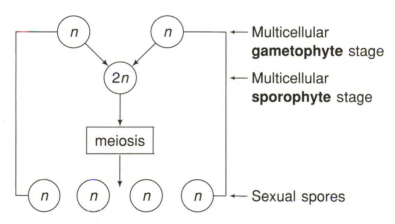

NOTE: The sporophyte stage is the most conspicuous in flowering plants and ferns, whereas the gametophyte stage is the most conspicuous in mosses.

7 8 9

21

11

Mendel's laws (8 9), in conjunction with random fertilization, generate standard genotypic and phenotypic ratios in the progeny.

Our examples show how this principle produces a $3:1$ ratio in a monohybrid cross, and a $9:3:3:1$ ratio in a dihybrid cross.

Example 1

Parents Aa × Aa (Monohybrid cross)

Random gametic pairing

	♂ $\frac{1}{2}$ A	$\frac{1}{2}$ a
♀		
$\frac{1}{2}$ A	AA	Aa
$\frac{1}{2}$ a	Aa	aa

Genotypic ratio
$\frac{1}{4}$ AA
$\frac{1}{2}$ Aa
$\frac{1}{4}$ aa

Phenotypic ratio
$\frac{3}{4}$ A−
$\frac{1}{4}$ aa

Example 2

Parents AaBb × AaBb (Dihybrid cross)

Random genetic pairings

♂ ⟍ ♀	$\frac{1}{4}$ AB	$\frac{1}{4}$ Ab	$\frac{1}{4}$ ab	$\frac{1}{4}$ aB
$\frac{1}{4}$ AB	AABB	AABb	AaBb	AaBB
$\frac{1}{4}$ Ab	AABb	AAbb	Aabb	AaBb
$\frac{1}{4}$ ab	AaBb	Aabb	aabb	aaBb
$\frac{1}{4}$ aB	AaBB	AaBb	aaBb	aaBB

Genotypic ratio

$$\frac{1}{16} : \frac{1}{16} : \frac{1}{16} : \frac{1}{16} : \frac{4}{16} : \frac{2}{16} : \frac{2}{16} : \frac{2}{16} : \frac{2}{16}$$

Phenotypic ratio

$\frac{9}{16}$ A–B–

$\frac{3}{16}$ A–bb

$\frac{3}{16}$ aaB–

$\frac{1}{16}$ aabb

12

In most organisms, sex or mating type are determined genetically.

Sex determination is by special genes, and these are either found on heteromorphic sex chromosomes (Examples 1–3) or on homomorphic sex chromosomes (Example 4).

Example 1 Humans

$$XX = Female; XY = Male$$

Presence or absence of Y determines sex.

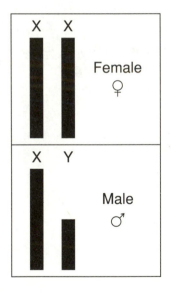

Example 2 *Drosophila*

$$XX = Female; XY = Male$$

Sex is determined by number of sex chromosomes in relation to the number autosomes.

Example 3 Birds

$$ZW = \text{Female; } WW = \text{Male}$$

Note the heterogametic sex is the female here.

Example 4 Yeast (n)

Mating type a is determined by gene MATa
Mating type α is determined by its allele MATα

The two mating types are not visibly different, but only an a and α cell can pair to constitute a diploid meiocyte.

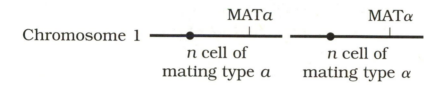

MATa MATα

Chromosome 1

n cell of n cell of
mating type a mating type α

13 14 61

13

Genes located on the sex chromosomes show sex-linked inheritance.

The genes that show sex-linked inheritance are mainly not concerned with sex determination.

Most cases of sex-linked inheritance are shown by genes on the region of the X that has no counterpart on the Y.

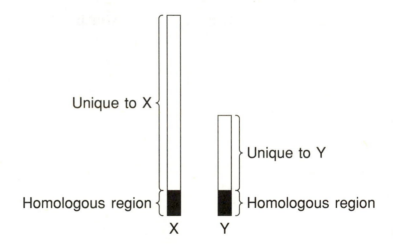

Genes on the unique region of the X are said to be hemizygous in males. **12 14 61**

Example   Inheritance of white eye in *Drosophila*

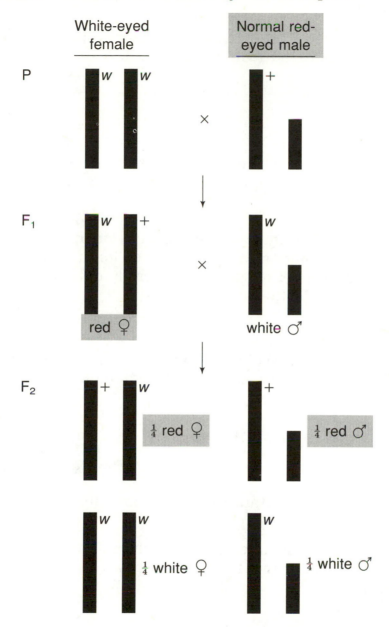

NOTE: +, the wild allele of *w*, is dominant.

27

14

The mode of inheritance of a human phenotype can be inferred from pedigree analysis.

Because controlled "crosses" cannot be made, inheritance must be deduced from matings that have already taken place.

NOTE: (a) Generally pedigrees concern phenotypes which are medically abnormal, and are therefore relatively rare; (b) The number of offspring is generally not large enough to reveal Mendelian ratios as in **11** .

SOME CLUES

1. The occurrence of affected individuals in every generation indicates dominant inheritance **6** .

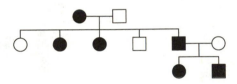

2. X-linked inheritance can be eliminated by male-to-male transmission.

The above pedigree is obviously autosomal because male-to-male transmission occurs.

Infer:

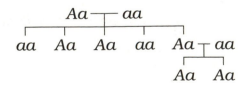

3. Occurrence of only affected males generally indicates X-linked recessive inheritance **13** .

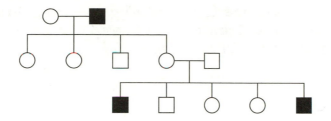

Infer:

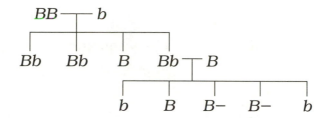

8 9 11

15

A chromosome is essentially one long double-stranded molecule of DNA, and the genes are regions on this molecule.

Average human chromosome:

~5 cm of DNA

Average fungal or bacterial chromosome:

~1 mm of DNA

There are proteins associated with the DNA: in prokaryotes these are **regulatory proteins;** in eukaryotes there are **analogous regulatory proteins** and large amounts of **histone protein,** which has organizational and regulatory properties. **70 72 73**

Example   Bacterial chromosome

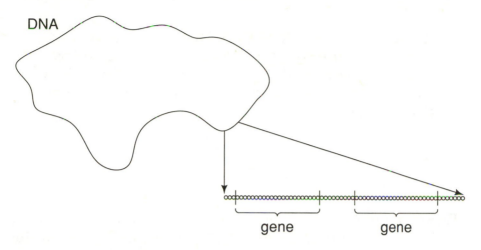

DNA

gene   gene

Example   Eukaryotic chromosome

A double strand of DNA is coiled around histone protein to form particles called **nucleosomes.** These are the components of **chromatin.**

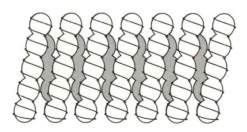

The nucleosome-carrying strand is then supercoiled.Other (nonhistone) proteins are associated with this chromatin structure.   **3**

16

A gene is a stretch of DNA that comprises a transcriptional unit. It consists of a sequence that is transcribed to a functional RNA product and regulatory sequences that enable transcription to occur.

This principle shows how hereditary *information* (genes) is converted into biological *form* (determined largely by proteins).

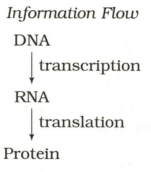

Information Flow

DNA

transcription

RNA

translation

Protein

A COMMONLY OBSERVED GENE STRUCTURE

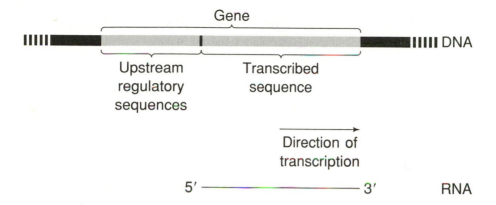

NOTE: For more details see **69** through **73** .

In nucleotide sequences **93** genes can often be found by identifying regulatory and transcription/translation initiation or termination sequences.

33

17

DNA is a double helix consisting of two complementary nucleotide strands of opposite polarity.

DOUBLE HELIX

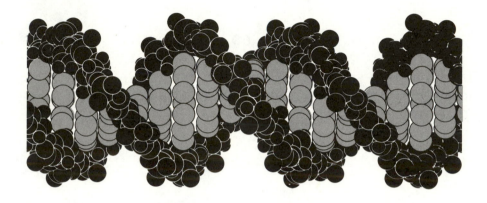

COMPLEMENTARITY

A always hydrogen bonds to T

G always hydrogen bonds to C

A typical sequence of DNA:

```
5'----C  T  A  G  C  T  A  T  T  A  C  G----3'
3'----G  A  T  C  G  A  T  A  A  T  G  C----5'
```

OPPOSITE POLARITY

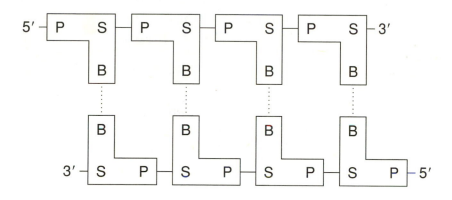

S = sugar   G = guanine

B = base (G, C, A, or T)   T = thymine

P = phosphate   C = cytosine

······ = hydrogen bond   A = adenine

⌐ = nucleotide

15 18 20

18

The template function afforded by base complementarity enables information transfer from DNA to DNA, from DNA to RNA and from RNA to protein.

This principle operates at cell division, when more DNA is made on DNA templates, and prior to protein synthesis when RNA is made from a DNA template.

DNA → DNA

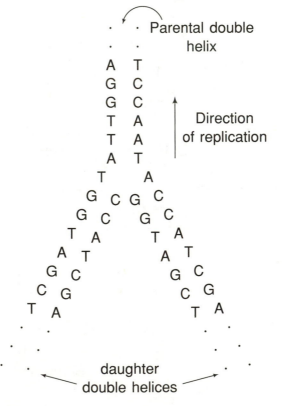

Refer to **17 20** .

DNA → RNA

```
5'----A  T  G  G  C  T  A  A  C----3'
3'----T  A  C  C  G  A  T  T  G----5' ←Template
                                        strand
```

mRNA

```
5'----A  U  G  C  C  U  A  A  C----3'
```

RNA → PROTEIN

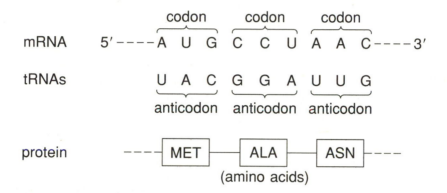

Refer also to **19** and **57** .

19

Genes exert their effects on the phenotypes of organisms by expressing a product that is either

- **A catalytic protein,**

- **A noncatalytic protein, or**

- **Nontranslated RNA.**

Genes interact with the environment to produce an organism: here we see that genes contribute mainly protein, but also certain crucial RNA types.

Examples

- Catalytic protein: e.g., enzyme

- Noncatalytic protein: e.g., keratin in hair or hemoglobin

- Nontranslated RNA: e.g., tRNA or rRNA

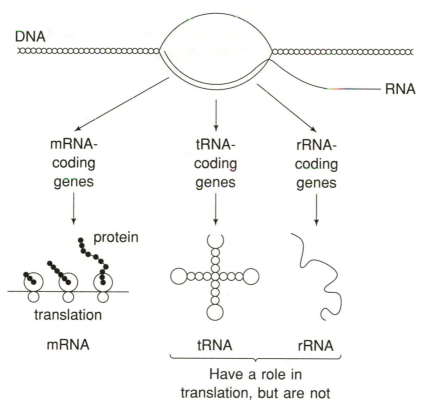

Transcription

DNA

RNA

mRNA-
coding
genes

tRNA-
coding
genes

rRNA-
coding
genes

protein

translation

mRNA

tRNA

rRNA

Have a role in
translation, but are not
themselves translated

18 57

20

DNA replicates semiconservatively, and bidirectionally in most organisms.

PROCESS OF REPLICATION

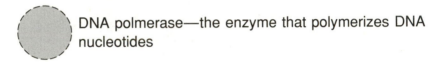

DNA polmerase—the enzyme that polymerizes DNA nucleotides

NOTE: growth of new strands is always 3' growth.

OVERALL EFFECT

Parental
double
helix

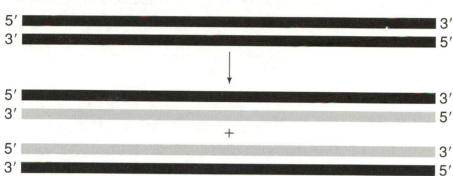

Daughter molecules each consist of template strand and one newly polymerized strand.   **17 18 78**

21

DNA and RNA are synthesized at 3′ growing sites: translation of mRNA is in the 5′ to 3′ direction, and polymerization of amino acids is at the carboxy terminus of the growing protein.

NOTE: this principle draws attention to the crucial role of polymers in the living process.

DNA SYNTHESIS

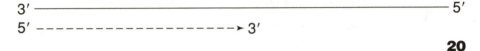

20

RNA SYNTHESIS

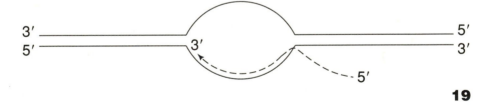

19

PROTEIN SYNTHESIS

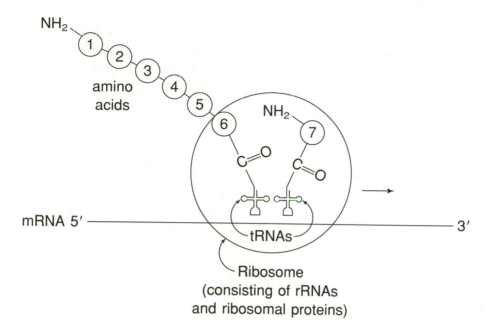

tRNAs provide specificity for individual amino acid attachment. Anticodons of tRNAs align with specific codons of mRNA **18** . Interaction with ribosomes and enzymatic components facilitate translation **57** .

22

A recombinant is a cell, descended from a diploid (or partially diploid) cell, that bears a new combination of the genes born by the two input genotypes for that diploid (or partial diploid).

All the following examples illustrate this coming together of genomes (or genomic fragments) to produce a novel genetic combination.

Example Meiotic recombination

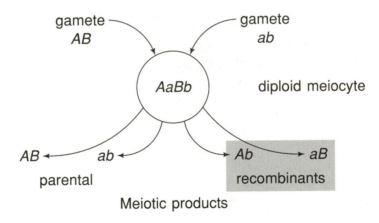

Meiotic products

Example   Bacterial conjugation

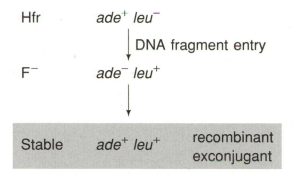

Example   Bacterial transduction

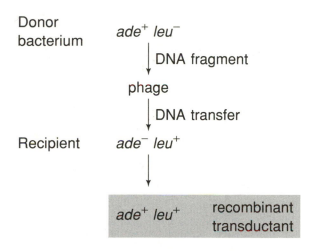

23 through **28 31 34 35 85**

Contrast with recombinant DNA **85** .

23

The study of gene transmission and recombination in bacteria is based on partial diploids (merodiploids)

There are at least three ways by which merodiploids can be formed.

1. *Conjugation*

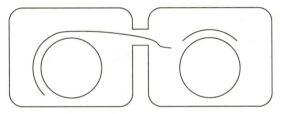

28

2. *Transduction*

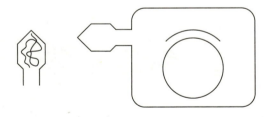

31 86

3. *Transformation*

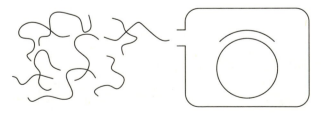

87

RECOMBINATION IN A MERODIPLOID

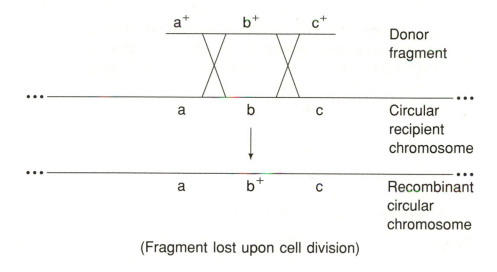

(Fragment lost upon cell division)

3 22

NOTE: with a circular recipient chromosome, recombination must be formally by a double crossover; otherwise the circle will be opened

47

24

In eukaryotes, recombinants can be produced by independent assortment of nonhomologous chromosomes or by homologous exchange. In prokaryotes and viruses recombination is by homologous exchange.

EUKARYOTIC RECOMBINATION

Independent Assortment

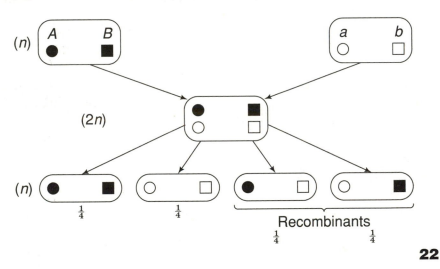

22

NOTE: 50 percent recombinants is characteristic of independent assortment and usually means the genes are on separate chromosomes.

Crossing Over

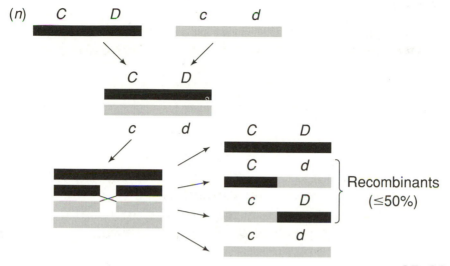

25 33

PROKARYOTIC RECOMBINATION

Bacterial merodiploid (partial diploid) produced in conjugation, transformation, or transduction:

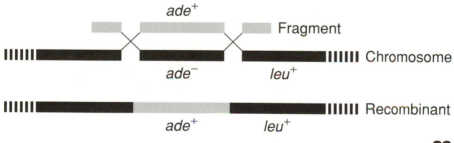

23

25

At meiosis, crossing over occurs between any two nonsister chromatids at the four-chromatid stage.

Proof The proof comes from the observation that when an *individual* meiosis is studied via tetrad analysis, a crossover in a genetically marked region results in four different meiotic product genotypes (not two). **32**

Single crossover occurring at the four-chromatid stage:

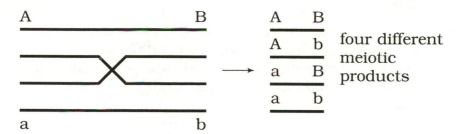

four different
meiotic
products

Single crossover occurring hypothetically at the two-chromosome stage:

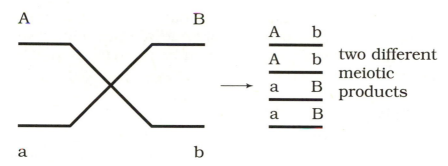

two different
meiotic
products

27

26

The recombinant frequency (RF) between loci on the same chromosome is roughly proportional to their distance apart, allowing the use of RF units as chromosomal mapping units.

This principle is based on the idea that homologous exchanges occur randomly along paired chromosomes, so two loci that are far apart provide a relatively greater opportunity for exchanges to occur between them than loci that are closer together.

Example In eukaryotes

1 map unit (centiMorgan—cM)
$\equiv$ 1 percent recombinant meiotic products

In a $2n$ test cross

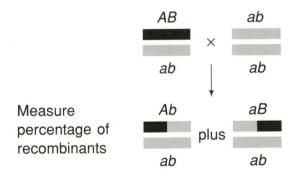

If RF = 10 percent, then map is:

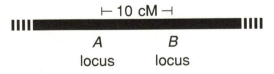

Example In a bacterial partial diploid during conjugation:

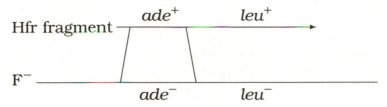

1. Select ade^+ exconjugants (hence know leu^+ entered too)

2. Measure percentage that is also leu^-. Map distance between ade and leu:

$$\text{Map distance} = \frac{ade^+\ leu^-}{\text{Total}\ ade^+}$$

23 28

Corollary In eukaryotes, the frequency of recombinants for linked genes never exceeds 50 percent.

The occurrence of a crossover in one chromosomal region is often influenced by the
occurrence of a crossover in an adjacent
region: this effect is called interference.

Example Calculation of interference in a standard
trihybrid ("3 point") test cross

$$AA\ BB\ CC \quad \times \quad aa\ bb\ cc$$

$$\downarrow$$

Aa Bb Cc	×	aa bb cc
Trihybrid		Tester

$$\downarrow$$

Gametes

1.	A	B	C	277
2.	a	b	c	277
3.	A	b	c	123
4.	a	B	C	123
5.	A	B	c	73
6.	a	b	C	73
7.	A	b	C	27
8.	a	B	c	27
				1000

Analysis Chromosomes of trihybrid must be

A	B	C
a	b	c

- Gamete types 1 and 2 are nonrecombinant.
- Types 3 and 4 arise from *single* crossover in the *A–B* region.
- Types 5 and 6 arise from *single* crossover in the *B–C* region.
- Types 7 and 8 are *double* crossovers.

Hence,

Rec Freq $A–B$ = (123 + 123 + 27 + 27) ÷ 1000 = 30 cM
Rec Freq $B–C$ = (73 + 73 + 27 + 27) ÷ 1000 = 20 cM

$$
\begin{array}{ccc}
A & B & C \\
\vdash & \vdash & \dashv \\
\text{30 cM} & \text{20 cM}
\end{array}
$$

$$Interference = 1 - \frac{\text{Observed double crossovers}}{\text{Expected double crossovers}}$$

$$= 1 - \frac{54}{(.3 \times .2 \times 1000)}$$

$$= 1 - \frac{54}{60}$$

$$= \frac{6}{60} = 10 \text{ percent}$$

22 24 25 26

28

In bacteria, the three basic chromosome mapping techniques are:

1. **Time of Hfr—high frequency of recombination (29)—gene entry into F⁻**

2. **Cotransduction frequency by phage**

3. **Recombinant frequency in partial diploids**

TECHNIQUE 1

Example

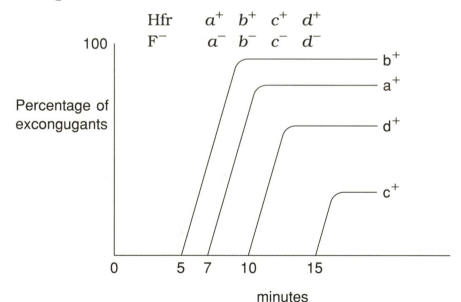

Hfr $\quad a^+ \quad b^+ \quad c^+ \quad d^+$

F⁻ $\quad a^- \quad b^- \quad c^- \quad d^-$

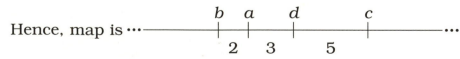

Hence, map is ···

TECHNIQUE 2

Example

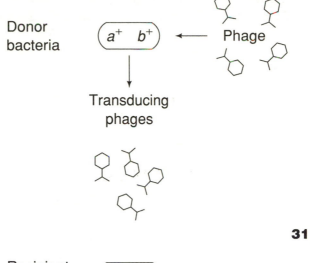

Donor bacteria $\boxed{a^+ \quad b^+}$ ⟵ Phage

Transducing phages

31

Recipient bacteria $\boxed{a^- \quad b^-}$

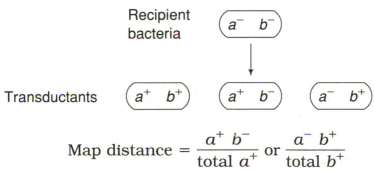

Transductants $\boxed{a^+ \quad b^+}$ $\boxed{a^+ \quad b^-}$ $\boxed{a^- \quad b^+}$

$$\text{Map distance} = \frac{a^+ \, b^-}{\text{total } a^+} \text{ or } \frac{a^- \, b^+}{\text{total } b^+}$$

TECHNIQUE 3 previously described, **23** .

29

Plasmids are circular or linear DNA molecules
that are found replicating independently of the
genomic DNA in bacteria and fungi.

Example Drug resistance (R) and sex (F) plasmids in *E. coli*

F⁺ cell

Hfr cell
(integrated F)

antibiotic
resistant cell
caused by R
plasmid

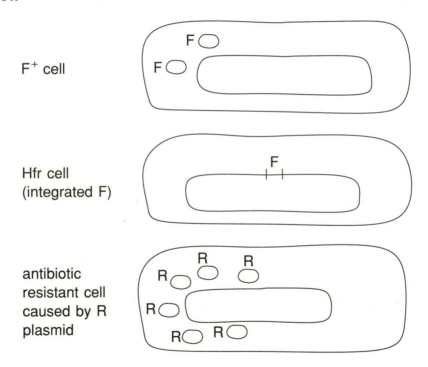

3 30 85 86

Example   Yeast 2μ plasmid

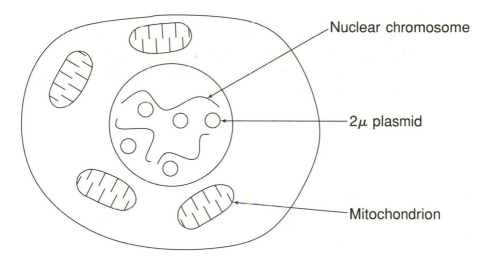

Example   Plant tumor-inducing (T_i) plasmid in *Agro-bacterium*. Part of its sequence (T) is transferred to an infected plant cell.

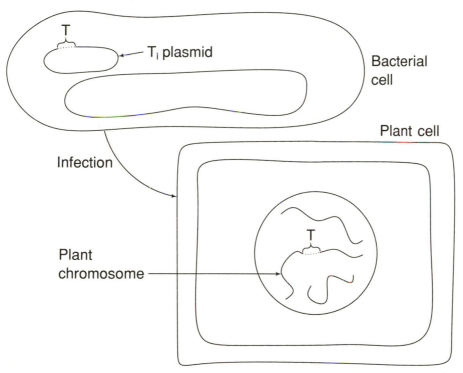

30

Plasmids can be inherited as autonomously replicating cytoplasmic/nucleoplasmic elements, or (when inserted) in the same manner as genomic elements.

Plasmids are extragenomic DNA molecules that can persist in a cell **29** . They are generally nonessential for most cell functions.

AUTONOMOUS

Example 1   F particles in *E. coli* **29**

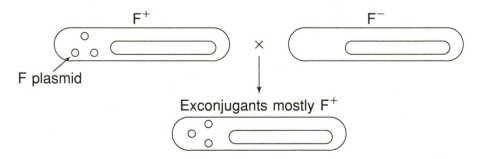

Example 2   Yeast 2μ plasmid

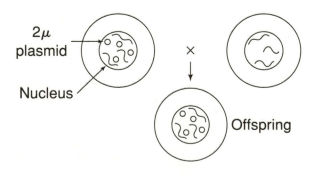

29

INSERTED

Example 1   F insertion to create Hfr

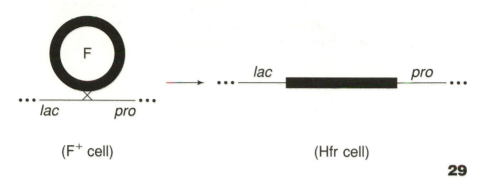

(F⁺ cell)   (Hfr cell)

29

Example 2   Yeast plasmid insertion can be followed if it carries a suitable yeast gene.

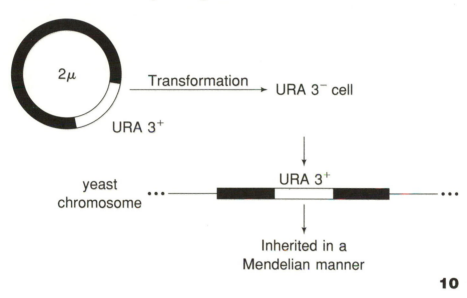

10

Example 3   T$_2$ plasmid, inserted into a plant chromosome, is inherited in a Mendelian manner.   **29**

31

The study of the transmission and recombination of phage genes is based on the coinfection of a bacterial culture by two different phage genotypes.

Hence the phage cross provides an opportunity for the union and reconstruction of different genomes, analogous to the sexual cycles of organisms.

Note that in such a situation there are:

1. Several types of coinfection

> phage 1 × phage 2 (shown on right)
> phage 1 × phage 1
> phage 2 × phage 2

2. Multiple rounds of infection and recombination.

Example

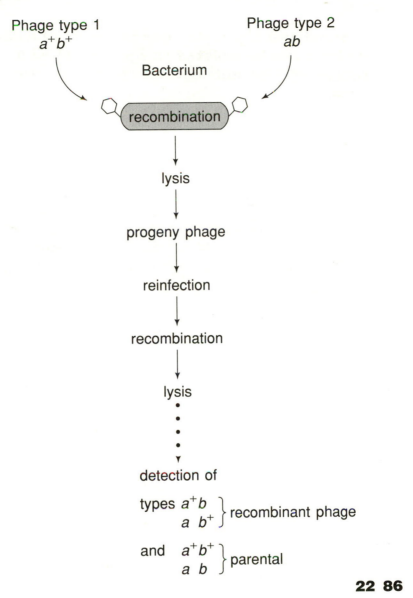

Phage type 1
a^+b^+

Phage type 2
ab

Bacterium

recombination

lysis

progeny phage

reinfection

recombination

lysis
•
•
•
•

detection of

types a^+b } recombinant phage
     $a\ b^+$

and a^+b^+ } parental
      $a\ b$

22 86

32

In fungi and algae, the four products of a single meiosis can be isolated as a group called a tetrad: this permits several specialized analyses.

Some of these analyses are:

1. Study of the processes of segregation and recombination of genes **8 9 24** at the level of single meiosis, rather than inferring them from the study of random meiotic products.

   For example, in a cross between linked genes:

   $$A B \times a b$$

   tetrads can be of three types

Parental ditype (PD)	Tetratype (T)	Nonparental ditype (NPD)
AB	AB	Ab
AB	Ab	Ab
ab	aB	aB
ab	ab	aB
(No crossovers, or two-chromatid double crossover)	(One crossover, or three-chromatid double crossover)	(Four-chromatid double crossover)

2. Calculation of gene–centromere distances

$\bigcirc = A$
$\bigcirc = a$

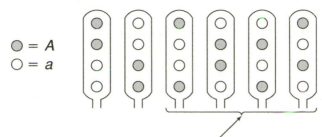

In these tetrads (M_{II} tetrads) a crossover has occurred between the gene and its centromere, e.g.,

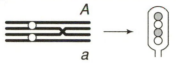

Gene–centromere distance (in centiMorgans) = frequency of M_{II} tetrads ÷ 2)

3. Study of mechanism and time of crossing over

25 33 34

4. Correction of map distances for double crossovers (which do not contribute proportionally to the production of recombinants).

Formula

Corrected map distance = 50 (T + 6 NPD)

22 26

33

Homologous DNA exchange involves a heteroduplex intermediate that resolves either into a recombinant or a nonrecombinant molecule.

In this figure the four chromatids paired in a eukaryote meiosis are represented as four DNA molecules.

Example   In eukaryotes

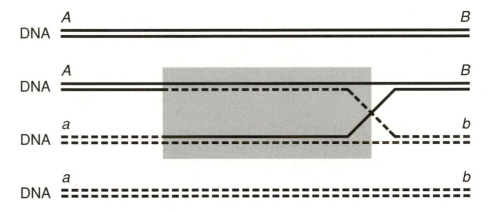

heteroduplex region

There are two possible resolutions of this intermediate:

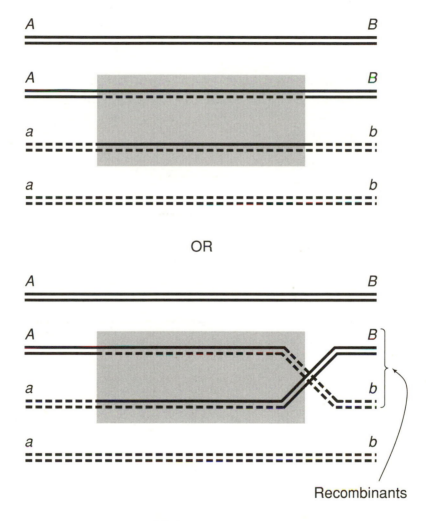

This is known as the **Holliday model** of crossing-over.

18 22 24 32 34

34

If a heteroduplex recombination intermediate spans a region of genetic heterozygosity, then correction of any mismatched bases can lead to gene conversion.

Gene conversion is the transfer of information from one DNA molecule to the same site in a homologous molecule.

CORRECTION

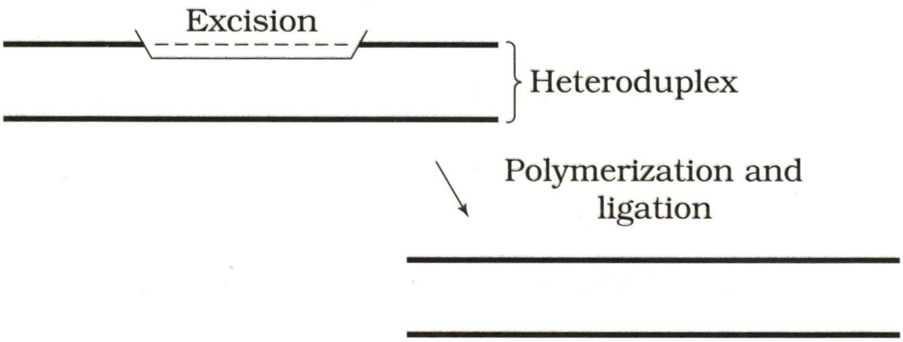

Excision

Heteroduplex

Polymerization and ligation

22 24 32 33

Example In fungal octads

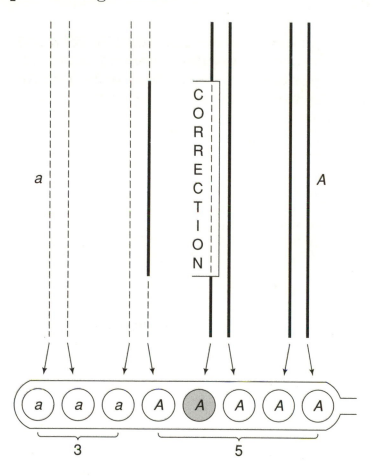

Aberrant ratio in ascus shows that gene conversion has occurred. Such ratios are rare because it is only rarely that heteroduplex and correction occur at a heterozygous locus.

Gene conversion can occur in diploid cells at meiosis or mitosis.

35

In diploid eukaryotic cells, mitotic recombination is a rare but regular occurrence: it can occur by mitotic crossing over or by mitotic chromosome assortment, and results in genotypic mosaics.

Example Mitotic crossing over

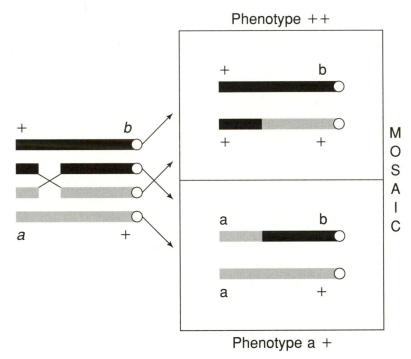

Phenotype ++

Phenotype a +

MOSAIC

NOTE: Alleles at the distal locus (*a*/+) have become homozygous.

22 24

Example Mitotic assortment by haploidization through chromosome loss

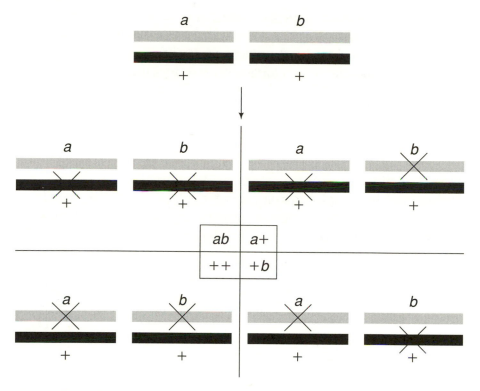

NOTE: See **36** for another haploidization-like process.

36

Human genes can be assigned to specific chromosomes by correlating retained human gene function and retained human chromosomes in man/rodent hybrid cells.

The technique is made possible by the unexplained loss of most human chromosomes from the hybrids, and by the ease with which human and rodent chromosomes can be distinguished under the microscope.

METHOD

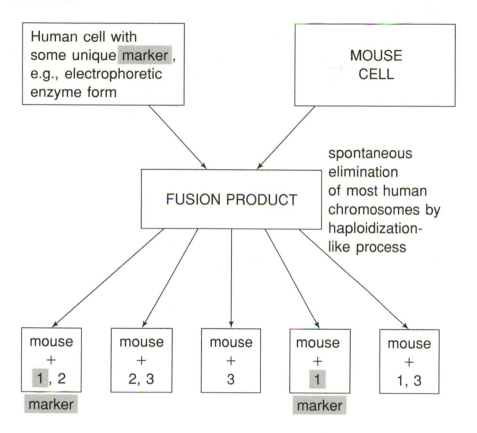

Gene for the marker must be on chromosome 1.

When human chromosomes bearing a series of deletions are used, the chromosomal locus of the marker can be pinpointed. **59 60**

37

In exerting their effects on the phenotype, genes do not act alone: they interact with each other and with the environment.

Example $c^h c^h$ in rabbits

At low temperatures

At high temperatures

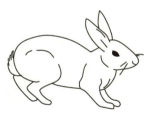

Example In metabolism, many genes cooperate to make the final products:

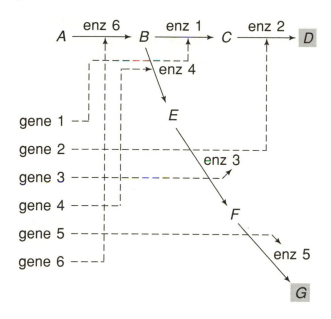

Hence, inactivation of one gene has ramifying effects—for example, gene 6 inactivation would lead to loss of product *D and G*. Genes often have such **pleiotropic** effects.

38 45

38

Gene interaction is revealed at the genetic level in eukaryotes by observing modified Mendelian ratios in standard crosses.

Different types of gene interaction give rise to different modified Mendelian ratios.

Example Gene S is a suppressor that causes a to become wild type in phenotype in a haploid eukaryote.

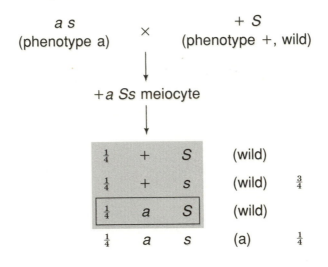

$$a\ s \qquad \times \qquad +\ S$$
(phenotype a)   (phenotype +, wild)

$\downarrow$

$+a\ Ss$ meiocyte

$\downarrow$

$\frac{1}{4}$	$+$	S	(wild)	
$\frac{1}{4}$	$+$	s	(wild)	$\frac{3}{4}$
$\frac{1}{4}$	a	S	(wild)	
$\frac{1}{4}$	a	s	(a)	$\frac{1}{4}$

11 37

Example In a diploid flowering plant, a pathway is:

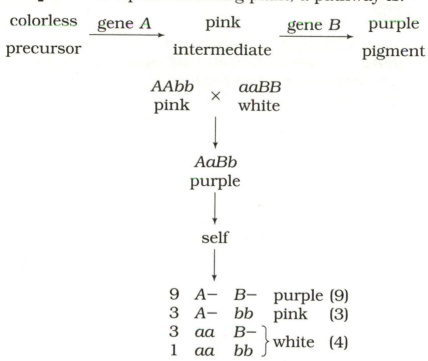

$$\begin{array}{ccc} \text{colorless} & \xrightarrow{\text{gene } A} & \text{pink} & \xrightarrow{\text{gene } B} & \text{purple} \\ \text{precursor} & & \text{intermediate} & & \text{pigment} \end{array}$$

AAbb $\times$ aaBB
pink   white

↓

AaBb
purple

↓

self

↓

9   A–   B–   purple (9)
3   A–   bb   pink   (3)
3   aa   B– ⎱
1   aa   bb ⎰ white   (4)

Other modified ratios possible:

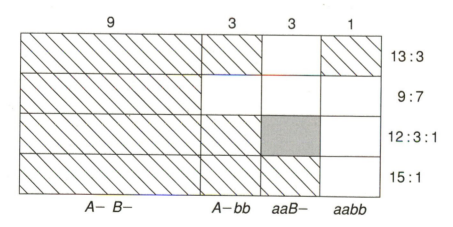

	9	3	3	1	
	A– B–	A–bb	aaB–	aabb	
					13:3
					9:7
					12:3:1
					15:1

39

Genes can change from one allelic form to another (gene mutation) and chromosome sets can change from one form to another (chromosome mutation).

Example Gene mutation

$$\text{Wild-type allele} \quad D^+ \xrightleftharpoons[\substack{\text{Reverse}\\\text{mutation}}]{\substack{\text{Forward}\\\text{mutation}}} D \quad \text{Dominant mutation}$$

$$\text{Wild-type allele} \quad r^+ \xrightleftharpoons[\substack{\text{Reverse}\\\text{mutation}}]{\substack{\text{Forward}\\\text{mutation}}} r \quad \text{Recessive mutation}$$

6

Example One kind of chromosomal mutation—a translocation

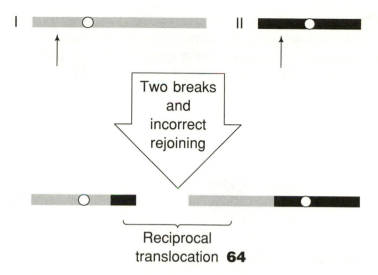

Two breaks
and
incorrect
rejoining

Reciprocal
translocation **64**

Mutation is the basis of all genetic variation.   **1**

40

Recessive mutations generally represent lack of, or reduced gene function. Dominant mutations can represent new gene function or interference with normal gene function.

The basis of this principle is that when a state of activity and a state of inactivity are united, the active state prevails.

RECESSIVE EXAMPLE

The recessive human disease phenylketonuria (PKU) is caused by low levels of the enzyme phenylalanine hydroxylase.

In a heterozygote p^+p

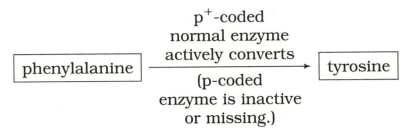

phenylalanine $\xrightarrow[\substack{\text{(p-coded} \\ \text{enzyme is inactive} \\ \text{or missing.)}}]{\substack{p^+\text{-coded} \\ \text{normal enzyme} \\ \text{actively converts}}}$ tyrosine

6

DOMINANT EXAMPLES

1. Nonsense suppressors in yeast are due to tRNA gene mutation in the anticodon region allowing translation of a nonsense (stop) codon.

In a diploid cell, *SS⁺ mm*

S^+-coded tRNA anticodon does not bind (e.g., GUC)

S-coded anticodon will bind

a.a.

AUC

UAG

mRNA

Nonsense codon in mutant

57

2. Mutant polypeptide distorts dimer in an M/+ heterozygote.

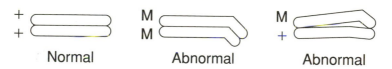

+ +   Normal   M M   Abnormal   M +   Abnormal

41

Mutation occurs spontaneously at low frequency or may be induced by the application of mutagens (certain chemicals and radiation).

Mutagens are routinely used experimentally in the search for specific mutations.

Some agents that are mutagenic in most organisms:

- Ethyl methane sulfonate (EMS)

- Nitrosoguanidine (MNNG)

- UV rays

- X rays

Most mutagens cause combinations of gene and chromosome mutation.

39

Typical dose response curves for gene mutation and translocations (an example of chromosome mutation) **64**

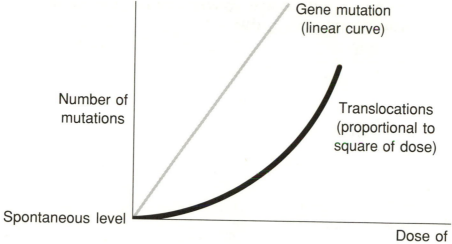

Number of mutations

Gene mutation (linear curve)

Translocations (proportional to square of dose)

Spontaneous level

Dose of mutagen

Some agents are very mutagenic, for example, in *Neurospora*, 25 μM nitrosoguanidine increases the number of mutations over spontaneous numbers by about *3000*-fold.

42

42

Some mutagens are specific in the types of mutations they induce.

Example Hydroxylamine

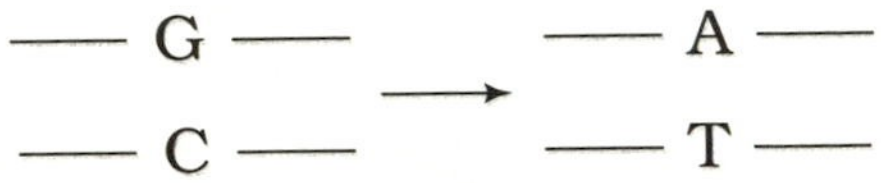

```
—— G ——          —— A ——
                ⟶
—— C ——          —— T ——
```

Example Nitrous acid

```
—— G ——          —— A ——
                ⇌
—— C ——          —— T ——
```

These base-pair substitutions result from specific chemical reactions with DNA.   **39**

Example Proflavine—adds or deletes nucleotide pairs.

$$\downarrow$$

—— CATCAT —— $\xrightarrow{+}$ —— CATCCAT ——
—— GTAGTA ——   —— GTAGGTA ——

$\searrow$ —— CATAT ——
—— GTATA ——

$$\downarrow$$

Example Ionizing radiation (e.g., X rays).

chromosome
break

51 59

43

In animals mutations occurring in germinal tissue may be transmitted to the next generation, but somatic mutations are never transmitted to the next generation. In plants somatic mutations in meristems may become part of germinal tissue.

The word *somatic* means of the body, whereas *germinal* refers to the tissue in which meiosis will occur.

PLANTS

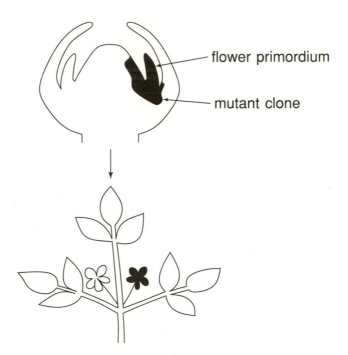

flower primordium

mutant clone

39

ANIMALS

Somatic

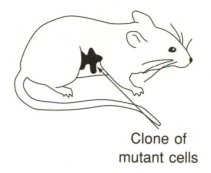

Clone of
mutant cells

46

Germinal

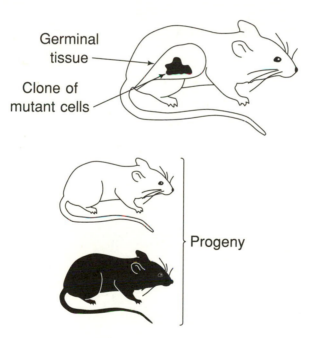

Germinal
tissue

Clone of
mutant cells

Progeny

The tendency of mutations to occur can be measured by the average number of mutation events per opportunity to occur (mutation rate) or by the proportion of mutants in a population (mutation frequency).

It is important to be able to measure the tendency of a gene to change, for example in testing the efficacy of mutagens.

MUTATION FREQUENCY

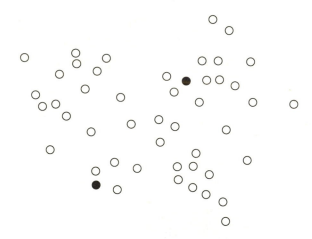

Frequency = $\frac{2}{50}$ (usually much lower)   **41 114**

88

MUTATION RATE

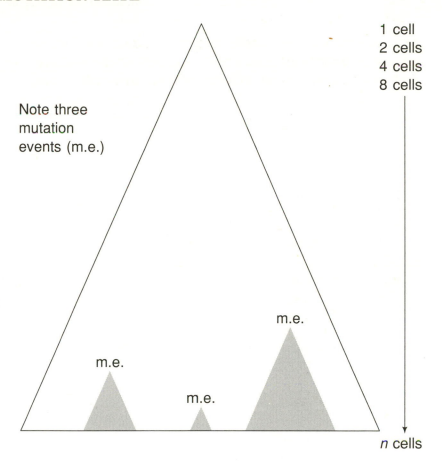

Hence, rate $= \dfrac{3}{n-1}$ mutation events per cell division (NOTE: to obtain n cells requires $n-1$ cell divisions). Both rates and frequencies generally fall in the range 10^{-5} to 10^{-8}.

45

Mutations can be used to dissect any biological process or structure: each mutant type identifies a separate component of the process.

This principle emphasizes the boundary of genetics. It is the use of spontaneous or induced genetic variants that sets genetics apart from other disciplines.

Example The study of a nucleotide sequence of a gene **93** becomes a genetic study when several variant sequences are compared at the functional level.

DNA sequence	Protein function
——— × ———————	normal
——————— × ————	abnormal
————————— × —	normal
— × ———————————	normal

possible
active site of
enzyme

× = altered nucleotide sequence.

Example In a *Neurospora* study, arginine-requiring mutants fell into four groups that defined four genes and their four enzymes involved in the biological synthesis of arginine.

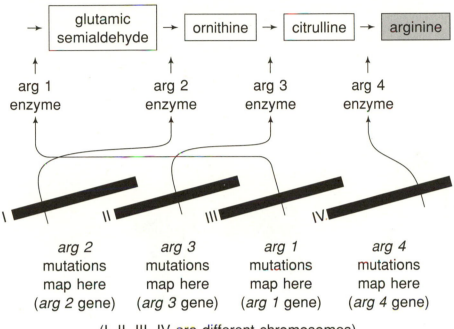

(I, II, III, IV are different chromosomes)

19 37

91

46

Genetic mosaics are rare individuals consisting of two or more genetically distinct cell lines.

A true mosaic must be distinguished from developmental patterning.

TRUE MOSAIC

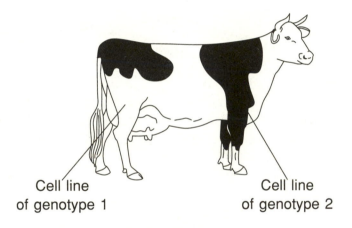

Cell line
of genotype 1

Cell line
of genotype 2

DEVELOPMENTAL PATTERNING

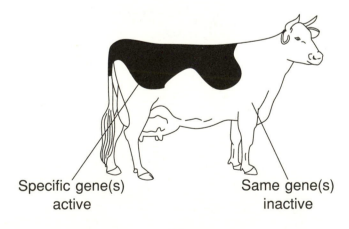

Specific gene(s)
active

Same gene(s)
inactive

ORIGINS OF MOSAICS

1. Somatic crossing over **35**

2. Somatic chromosome nondisjunction, e.g., in humans

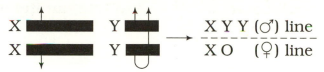

$$
\begin{array}{ll}
\text{X} & \text{Y} \\
\text{X} & \text{Y}
\end{array}
\longrightarrow
\begin{array}{l}
\text{X Y Y } (\male)\ \text{line} \\
\overline{\text{X O}\quad (\female)\ \text{line}}
\end{array}
$$

3. X-chromosome inactivation **61**

4. Fused zygotes

5. Organelle segregation **100**

6. Somatic mutation by DNA change **43**

7. Somatic mutation by transposon movement **58**

47

Mutant alleles can be used as genetic markers to keep track of a gene, a chromosomal region, a chromosome, a cell or cell line, an individual, or a population.

This is another principle that focuses on the usefulness of genetics in biological sciences.

Example *A chromosome* (see Mendel's laws **8 9**)

Example *A chromosome region* (see crossing over **24 26 27**)

Example A *cell line*—a genetic mosaic allows the use of a genetic marker to follow development. **46**

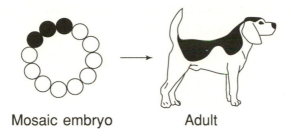

Mosaic embryo   Adult

Example A *population*—migration can be tracked using blood-group alleles as genetic markers. **109**

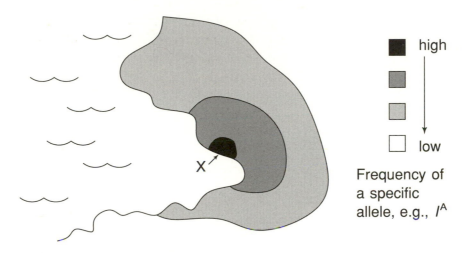

high

low

Frequency of a specific allele, e.g., I^A

Infer: migrants gained coast at *X* and genes spread through local population.

Example A length of DNA. **83**

48

Conditional mutations are particularly useful in genetic dissection 45 ; they permit

- **deleterious mutations to survive, and**

- **the determination of the developmental period of action of the gene.**

Many mutations are in genes of such crucial function that they are inherently lethal under normal conditions. In such cases conditional mutations that allow survival under normal conditions, and become mutant only under some extreme regime, are very useful.

Example Temperature-sensitive lethals, ℓ^{ts}—mutations of a crucial gene.

$\ell^{ts}\,\ell^{ts}$ fly

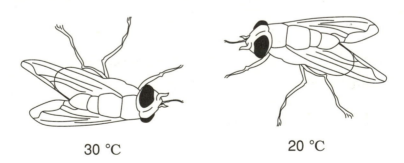

30 °C   20 °C

Shift-up experiments determine the temperature-sensitive period.

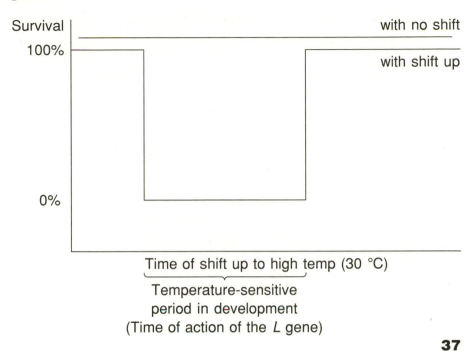

Survival

100%

0%

with no shift

with shift up

Time of shift up to high temp (30 °C)

Temperature-sensitive
period in development
(Time of action of the *L* gene)

37

97

49

Cells are equipped with enzymes that can repair DNA damage.

Organisms have survived since the origin of life in environments that are mutagenic; furthermore the basic DNA-processing systems of the cell often make errors. Hence the need for efficient repair.

Example Ultraviolet radiation damage can be repaired in several ways

a thymine
dimer

—— $\widehat{TT}$ ——
—— AA ——

(The main effect of UV rays on DNA)

Dimer Cleavage

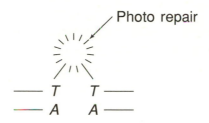

Excision Repair

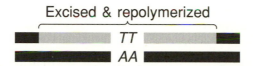

Dimer Bypass

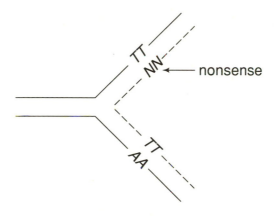

This method is highly error prone, so mutations occur.

NOTE: In humans, xeroderma pigmentosum is a disease in which there is genetic predisposition to skin cancer caused by a deficiency of the UV damage-repair system. Genotype is *rr* where *R* would provide normal repair functions.

39 51

50

The efficient study of mutations* requires a detection and a selection system.

Most mutations are rare, so when seeking mutations it is advantageous to design the system so that it provides maximum opportunity for the mutations to be obtained.

Example Recessive mutations in a diploid organism **40**

Mutagenize

AA BB CC DD, cross with *aa bb cc dd*

detect *aa* or *bb* or *cc* or *dd*
phenotype in progeny

(Must be due to mutation events
$A \rightarrow a$ or
$B \rightarrow b$ or
$C \rightarrow c$ or
$D \rightarrow d$.)

*For example, in genetic dissection see **45**

Example To select antibiotic-resistance mutations in yeast, culture on an antibiotic plate. Only resistant cells can form colonies.

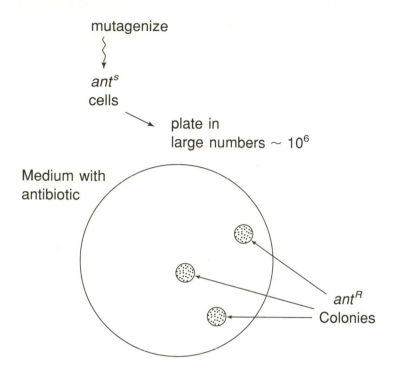

Success in this endeavour depends on whether an efficient selective system can be devised for the mutations you are interested in **39** .

$$(ant^s = \text{sensitive to antibiotic}$$
$$ant^R = \text{resistant to antibiotic.})$$

At the DNA level, the changes associated with mutations in a gene can be

1. Nucleotide pair substitution,

2. Nucleotide pair addition or deletion,

3. Larger additions or deletions, or

4. Transposon (Tn) insertion.

Any point mutation that arises could be from one of these four mechanisms, and further tests are required to determine which.

Example 1

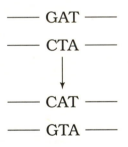

18 19

Example 2

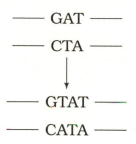

——— GAT ———

——— CTA ———

↓

——— GTAT ———

——— CATA ———

52

Example 3

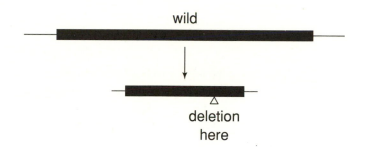

wild

deletion
here

59

Example 4

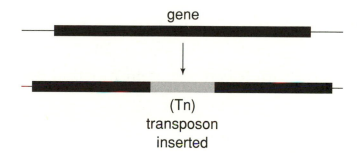

gene

(Tn)
transposon
inserted

58

52

At the protein level, the changes associated with gene mutations can be

1. **missense,**

2. **chain terminating,**

3. **silent, or**

4. **frame-shift.**

In determining the functional consequences of any mutation, these four possibilities need to be considered.

Example 1

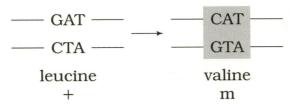

$$\text{—— GAT ——} \longrightarrow \text{—— CAT ——}$$
$$\text{—— CTA ——} \qquad \text{—— GTA ——}$$

leucine   valine
+   m

This can result in defective protein function **18 19 39** .

Example 2

glycine   stop
+   m

Example 3

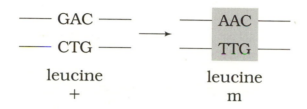

leucine   leucine
+   m

Example 4

───────────────► Reading direction

+ ··· − | T A A | A G C | T T C | G T T | A C G | − ···
   √   √   √   √   √

m ··· − | T A A | A C T | T C G | T T A | C G | − ···
   √   x   x   x   x

Deletion of G has shifted reading frame and all codons are incorrect downstream.   **51**

53

Complementation is the production of a wild-type phenotype when two mutant genomes occur in a common nucleus or cytoplasm.

Complementation is essentially a process in which genes at separate loci help each other out.

Using m_1 and m_2 as different genes

Example   A fungal heterokaryon

$$\frac{m_1 m_2^+}{} \qquad \frac{m_1^+ m_2}{}$$

      (n) (n)

54 89 101

Example In diploids

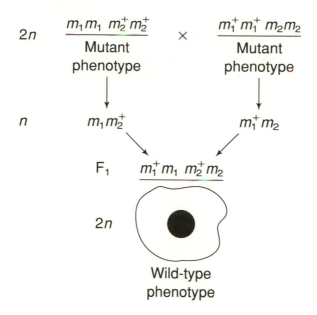

$$2n \quad \frac{m_1 m_1 \ m_2^+ m_2^+}{\text{Mutant phenotype}} \quad \times \quad \frac{m_1^+ m_1^+ \ m_2 m_2}{\text{Mutant phenotype}}$$

$$n \qquad m_1 m_2^+ \qquad\qquad m_1^+ m_2$$

$$F_1 \qquad m_1^+ m_1 \ m_2^+ m_2$$

2n

Wild-type
phenotype

Example In phages

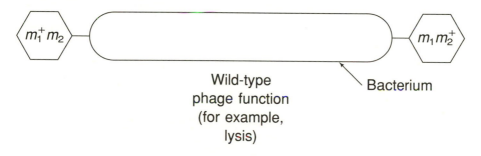

$m_1^+ m_2$

$m_1 m_2^+$

Wild-type
phage function
(for example,
lysis)

Bacterium

54

Recessive mutations that do not complement each other are said to be part of the same cistron. In general, a cistron can be equated with a gene.

The complementation test is often a useful alternative to the allelism test **5** in allocating new mutations to gene loci.

This principle provides a functional test for mutations within one gene/cistron.

Example The hypothetical mutations m_1, m_2, and m_3 all map to one approximate locus on a chromosome.

Investigate all possible pairwise combinations of mutants to test complementarity **53** .

RESULTS—Complementation

	m_1	m_2	m_3
m_1	–	+	+
m_2	+	–	–
m_3	+	–	–

CONCLUDE

- m_1 in one gene

- m_2 and m_3 in another closely linked cistron (**gene**), giving similar mutant phenotype.

For example,

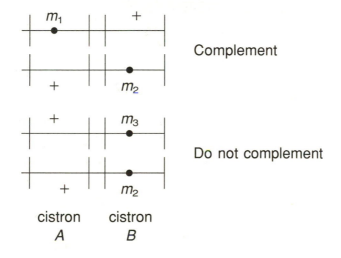

cistron   cistron
A   B

71

The position of mutant sites within a gene can be located by

1. **Intragenic recombination,**

2. **Amino acid analysis of mutant protein, or**

3. **Nucleotide sequencing of mutant gene.**

Knowledge of where a mutant allele is altered is useful in deducing the functional subregions of genes.

INTRAGENIC RECOMBINATION

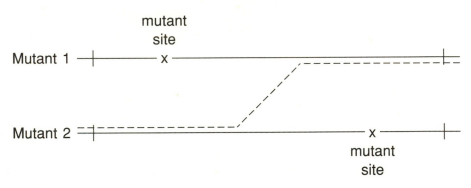

Map-using frequency of rare wild-type recombinants (--): this is proportional to distance between the mutant sites **22** .

AMINO ACID SEQUENCING

- Extract gene's protein product
- Compare mutant with wild-type sequence

NUCLEOTIDE SEQUENCING 93

- Clone gene **85** through **89**
- Compare mutant with wild-type sequence

+ [   ] DNA

m [■   ] DNA

↑
Single difference
determines mutant site

56

Spontaneous and induced mutant sites occur throughout the gene, but site-specific mutagenesis produces a change at a predetermined site.

Hence, rather than wait for the right mutation to occur by chance, or make do with those that are obtained, a specific change can be tailored to suit the question being asked concerning the gene's function.

Example

1. Sequence gene in question

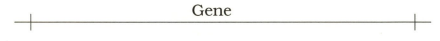

Gene

2. Select site for mutation

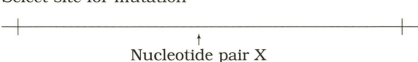

Nucleotide pair X

55

3. Chemically synthesize an oligonucleotide (~10 bp) containing desired base change

4. Hybridize oligo to single-stranded wild-type DNA, e.g., for phage M13 which has a single-stranded circular DNA genome.

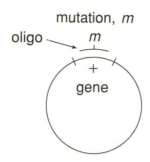

18

5. Synthesize a complementary strand using oligo as primer.

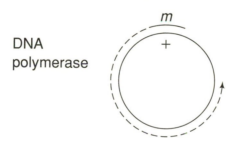

20

6. Transform recipients and select mutant phenotype.

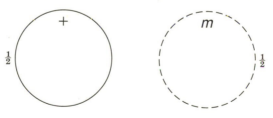

87

57

A suppressor mutation counteracts the pheno-typic effect of a prior mutation at a different site.

Detecting new mutations that reverse or cancel the effect of a prior mutation is an effective way of studying genetic interaction. Usually such suppressors are obtained in screens for *reversion* to wild type.

INTRAGENIC SUPPRESSOR

Example Original mutation: frame-shift (+1 bp)

Original
mutation

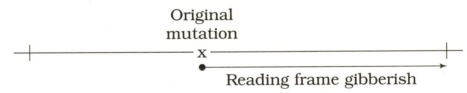

Reading frame gibberish

Suppressor mutation frame-shift (−1 bp)

original   suppressor

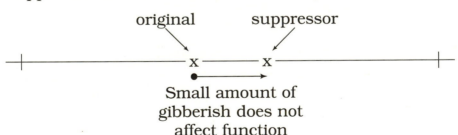

Small amount of
gibberish does not
affect function

51

114

EXTRAGENIC SUPPRESSOR

Example Nonsense suppression by tRNA suppressor **38**

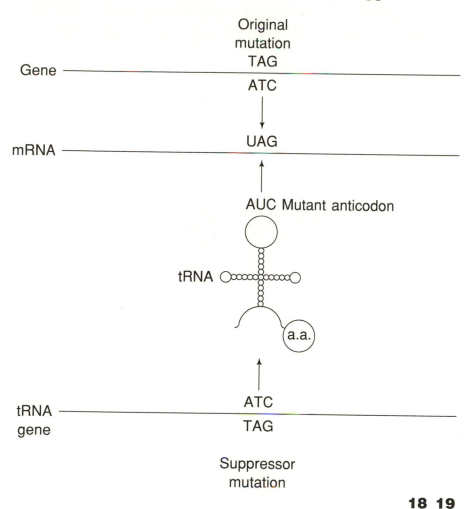

58

Certain kinds of DNA (called transposons) are mobile and can move throughout the genome.

Transposons have been found to be common in many organisms, but their movement can be detected only rarely.

The following are typical characteristics of transposons:

1. They have terminal repeats that bracket the moveable segment.

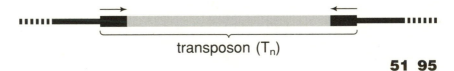

transposon (T_n)

51 95

2. They produce a mutant phenotype by inactivating a gene they enter.

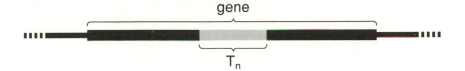

3. In some cases they produce unstable mutations because they are frequently excised.

Example

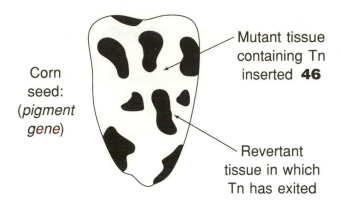

Corn seed: (*pigment gene*)

Mutant tissue containing Tn inserted **46**

Revertant tissue in which Tn has exited

4. They produce rearrangements at the site of transposition **59** through **64** .

117

59

A deletion of a chromosomal region can lead to genomic imbalance and/or expression of recessive genes on the other homolog.

Imbalance of gene dosage in a deletion heterozygote can be deleterious to the organism.

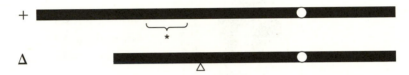

*This region is represented once in the homologous pair; other regions are represented twice.

Smaller deletions are tolerated better.

PSEUDODOMINANCE

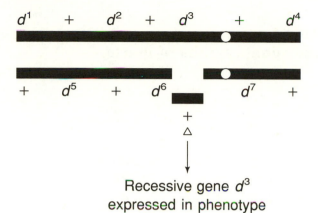

Recessive gene d^3
expressed in phenotype

40 60

119

60

Deletions can be used as probes to locate genes or functional regions of genes.

Example A restriction cut **81** followed by different degrees of erosion by a special exonuclease can be used to define functional promoter regions **72**

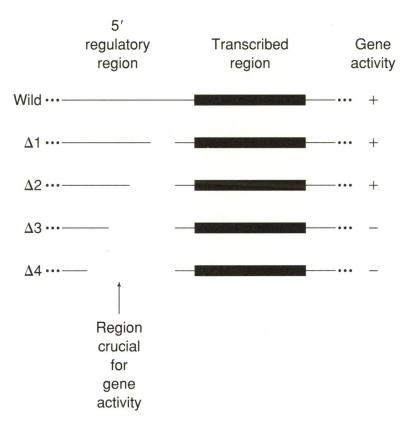

Example   Gene location by pseudodominance **59**

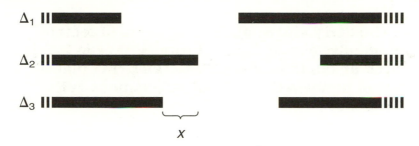

Δ_1

Δ_2

Δ_3

x

If a recessive mutant is expressed when in combination with Δ_1 and Δ_3 but not Δ_2, the mutation must be in region x.

Example   Gene location by recombination

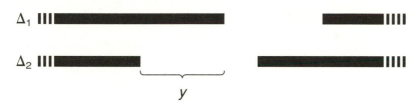

Δ_1

Δ_2

y

If a mutant can generate wild type recombinants with Δ_1 but not with Δ_2, the mutation must be in region y.

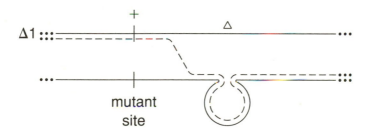

$\Delta 1$ $+$ Δ

mutant
site

61

In the early stages of development of female mammals, one of the X chromosomes is inactivated at random in each cell, and this inactivation persists through all subsequent cell divisions.

XX CELL

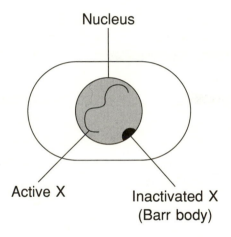

Nucleus

Active X

Inactivated X
(Barr body)

12 13

Consequence Adult female mammals are mosaics for the activity of all X-chromosome loci that are heterozygous.

Example Assume heterozygous for three loci

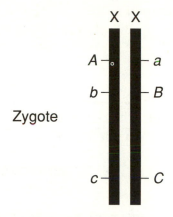

Zygote

Adult is a mosaic of cells of

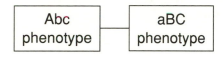

| Abc phenotype | | aBC phenotype |

46

62

Duplications can upset the chromosomal balance to produce general inviability or some unique phenotype.

Example The unique bar-eye phenotype in *Drosphila* is caused by a tandem duplication.

 Dup.

\+

The bar-eye determinant is dominant in combination with a wild-type chromosome.

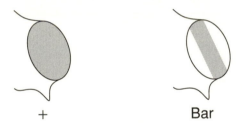

\+   Bar

SOME CAUSES OF DUPLICATIONS

1. Unequal crossing over **59**

Dup.

Δ

2. Transposon excision **58**

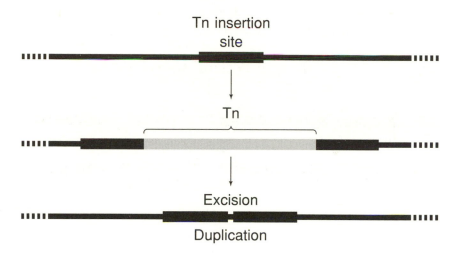

Tn insertion
site

Tn

Excision

Duplication

63

Heterozygous inversions show reduced local recombination and partial sterility.

An **inversion heterozygote** is composed of a normal chromosome in combination with an inverted chromosome.

Example   Paracentric inversion heterozygote

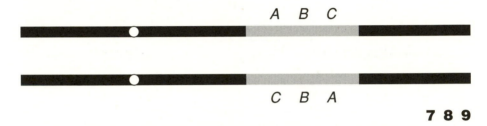

A   B   C

C   B   A

7 8 9

Pairing and crossover

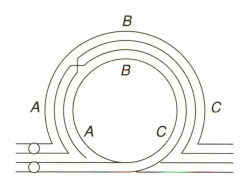

Chromosome segregation

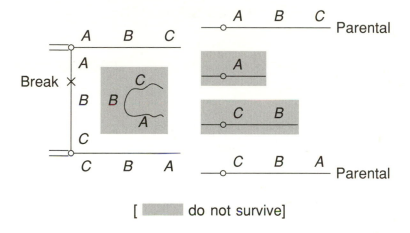

[▨▨▨ do not survive]

Hence, only parental genotypes survive, and crossover products are lost.

64

Heterozygous translocations can show linkage of genes on separate chromosomes and semisterility.

As in the case of inversions, these features are a result of the pairing conformation at meiosis.

Example Reciprocal translocation

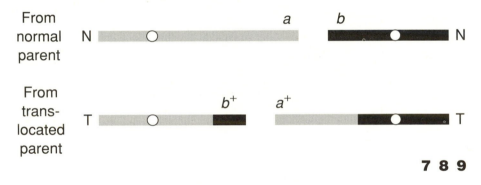

From normal parent

From trans-located parent

7 8 9

128

Pairing

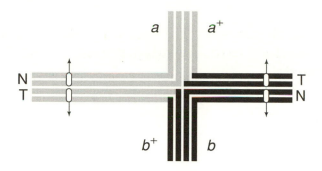

Segregation

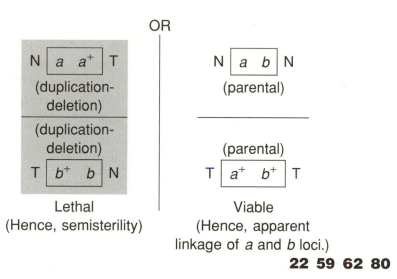

OR

| N | a a^+ | T |
(duplication-deletion)

(duplication-deletion)

| T | b^+ b | N |

Lethal
(Hence, semisterility)

| N | a b | N |
(parental)

(parental)

| T | a^+ b^+ | T |

Viable
(Hence, apparent
linkage of a and b loci.)

22 59 62 80

65

Aneuploids $(2n + 1, 2n - 1, 2n - 2, n + 1$ etc.) result mainly from nondisjunction at meiosis or mitosis.

Example Nondisjunction of human chromosome at first meiotic division

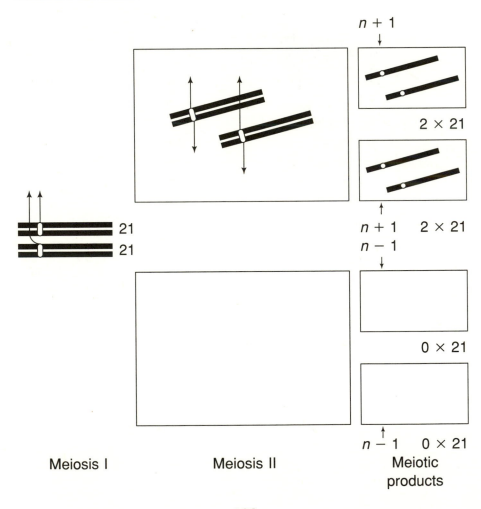

Down's syndrome [trisomy 21 (3 × 21)] results from union of a 2 × 21 gamete and a normal 1 × 21 gamete.

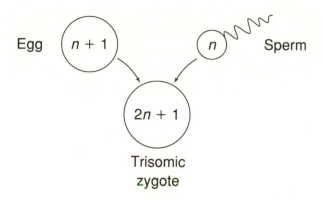

Egg (*n* + 1)   (*n*) Sperm

(2*n* + 1)

Trisomic
zygote

10

Example Aneuploid created at mitosis

XY   X ──────── ───── Y   XYY
zygote X ──────── ───── Y   ─────   Sexual   ♂
                                        mosaic
                                XO   ♀

7 46

66

Polyploids (3*n*, 4*n*, 5*n*, 6*n*, etc.) in plants can originate as autopolyploids of allopolyploids. They are usually more vigorous than diploids, but often exhibit reduced fertility because of irregular meiotic chromosome pairing.

AUTOPOLYPLOID

$$2n \xrightarrow[\text{(e.g., Colchicine)}]{\substack{\text{spindle poison} \\ \text{at mitosis}}} \begin{array}{c}\text{one } 4n \text{ cell} \\ \text{(instead of} \\ \text{two } 2n \text{ cells)}\end{array}$$

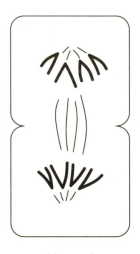

Normal

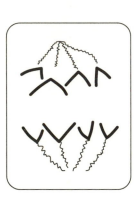

Spindle poisoned

7

ALLOPOLYPLOID

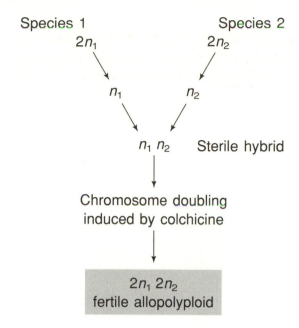

Species 1
$2n_1$

Species 2
$2n_2$

n_1 n_2

$n_1 n_2$ Sterile hybrid

Chromosome doubling
induced by colchicine

$2n_1 2n_2$
fertile allopolyploid

REDUCED FERTILITY

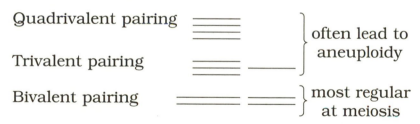

Quadrivalent pairing
⎫
⎬ often lead to
⎭ aneuploidy

Trivalent pairing

Bivalent pairing
⎫ most regular
⎬ at meiosis

67

Aneuploids and polyploids that are heterozygous for a genetic marker will produce characteristic phenotypic ratios.

Example Bivalent pairing in a tetraploid **66**

$$
\begin{array}{cccc}
A & A & a & a \\
1 & 2 & 3 & 4
\end{array}
$$

Possible pairings:

$$
1 \frac{A}{} 3 \frac{a}{} \quad \Big| \quad 1 \frac{A}{} 2 \frac{A}{} \quad \Big| \quad 1 \frac{A}{} 2 \frac{A}{}
$$
$$
2 \underset{A}{\phantom{\frac{}{}}} 4 \underset{a}{\phantom{\frac{}{}}} \quad \Big| \quad 3 \underset{a}{\phantom{\frac{}{}}} 4 \underset{a}{\phantom{\frac{}{}}} \quad \Big| \quad 4 \underset{a}{\phantom{\frac{}{}}} 3 \underset{a}{\phantom{\frac{}{}}}
$$

Results in meiotic products:

$$\tfrac{1}{6} \, AA$$
$$\tfrac{4}{6} \, Aa$$
$$\tfrac{1}{6} \, aa$$

$$\text{Upon selfing } \left(\frac{1}{6}\right)^2 = \frac{1}{36} \text{ will be } aaaa$$

$$\text{and } \frac{35}{36} \text{ will be } A\text{---}$$

In the case of aneuploids, the abnormal ratios are useful in associating mutations with specific chromosomes.

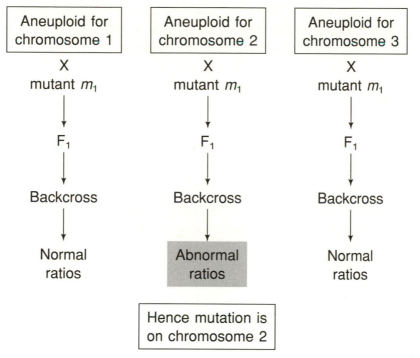

68

Propagation of higher animal and plant cells in vitro allows microbe-like somatic-cell genetic techniques to be applied to higher organisms.

This allows the resolving power of cell-plating techniques to be applied to higher eukaryotes.

Example Chinese hamster ovary (CHO) cells can be plated on medium to select for drug resistant mutants.

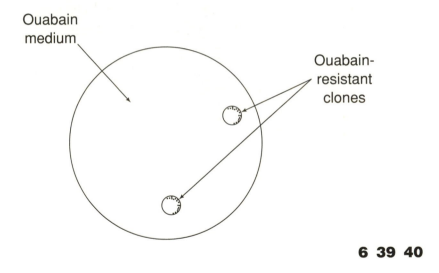

Ouabain medium

Ouabain-resistant clones

6 39 40

Example Haploid cells from plant anthers can be grown into haploid plants

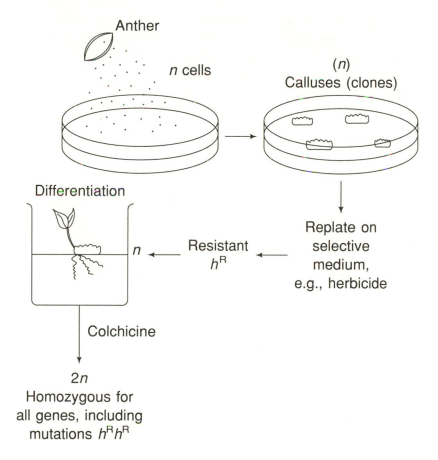

Thus, colchicine is used to generate a herbicide-resistant homozygous diploid ($2n$) from the haploid (n) plant.

69

The genomic DNA of different somatic cell types of an organism is the same, but differences occur in the genes that are expressed (from cell to cell and at different developmental stages).

This principle accounts for the range of different cell types in an organism, all derived mitotically from a single original cell (the zygote).

Example Muscle cell

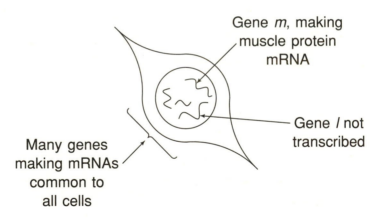

Gene *m*, making muscle protein mRNA

Gene *l* not transcribed

Many genes making mRNAs common to all cells

NOTE: Specialized DNA changes do, however, occur in cells producing immunoglobulins.

Example Lens cell of eye

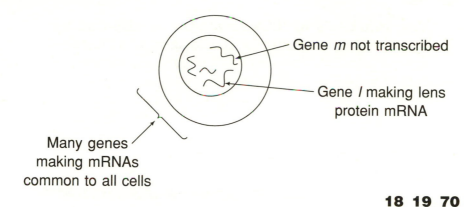

Gene *m* not transcribed

Gene *l* making lens
protein mRNA

Many genes
making mRNAs
common to all cells

18 19 70

70

The synthesis of RNA by a gene is affected by the specific regulatory sequences of the gene.

Example 1 Prokaryotic structural gene

| cis-acting regulatory region*: affects adjacent transcribed region only. | ↑ Site of transcription initiation | Transcribed to mRNA by RNA polymerase |

gene (or operon) **71**

Example 2 Eukaryotic structural gene (Transcribed by RNA Pol II)

| cis-acting regulatory region*: affects adjacent transcribed region only. | ↑ Site of transcription initiation | Transcribed to HnRNA** |

*Includes promoter **72** .
Heterogeneous nuclear RNA (unprocessed) **76 .

Example 3 Eukaryotic structural gene with an enhancer sequence.

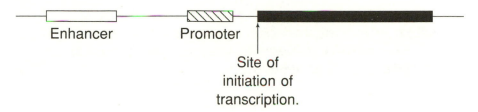

Enhancers are sequences that cause high levels of transcription in the adjacent gene. They may be located close or distant, and 5′ or 3′.

Example 4 Eukaryotic tRNA gene (Transcribed by RNA Pol III)

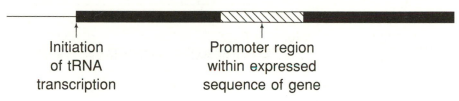

NOTE: Structural genes (prokaryotic and eukaryotic) encode protein products.

Mutations in regulatory regions are identified by their *cis action.* (i.e., they affect adjacent transcribed regions only, and not those of another homologue.)

In bacteria, genes with related functions are often adjacent and their proteins are translated off one mRNA molecule: this transcriptional unit is called an operon.

Example *Lac* operon in *E. coli*

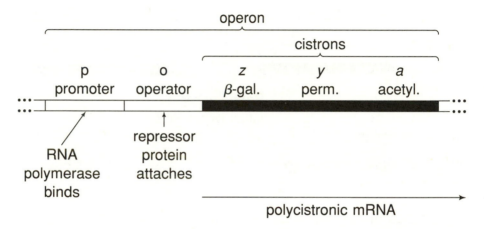

The cistrons *z*, *y*, and *a* encode the enzymes β galactosidase, permease, and transacetylase, respectively. The enzymes are all involved in the utilization of lactose by *E. coli*. They are coordinately regulated **73** .

Example Histidine operon in *Salmonella*

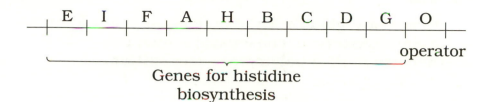

| E | I | F | A | H | B | C | D | G | O |

operator

Genes for histidine
biosynthesis

These genes control the following reactions

○ →^{G} ○ →^{E} ○ →^{I} ○ →^{A} ○ →^{H} ○ →^{F} ○ →^{B} ○ →^{C} ○ →^{B} ○ →^{D} histidine

Biosynthetic
pathway

Bacterial-style operons have not been identified in eukaryotes.   **74**

72

Promoter regions of similar types of genes have certain conserved (consensus) sequences involved in RNA polymerase recognition.

Consensus sequences are deduced from comparison of many genes, and their functionality is deduced by mutational analysis.

Example *E. coli* structural genes

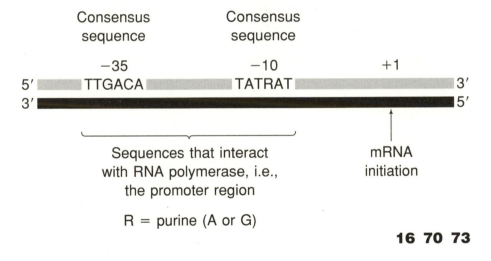

Sequences that interact with RNA polymerase, i.e., the promoter region

mRNA initiation

R = purine (A or G)

16 70 73

144

Example Eukaryotic structural genes

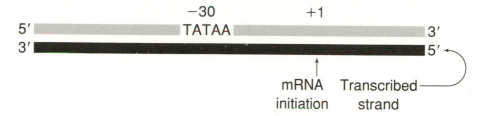

The TATAA sequence and its specific position in the eukaryotic gene direct the initiation site of RNA synthesis (i.e., at 30 nucleotides downstream of TATAA).

Upstream regulatory regions

70

145

73

Inducer and repressor functions are involved in the regulation of expression of some structural genes.

Inducers turn a gene on, and repressors turn it off.

Example The *lac* operon

Repressor molecule bound to DNA receptor site (operator) and removed by inducer (allolactose)

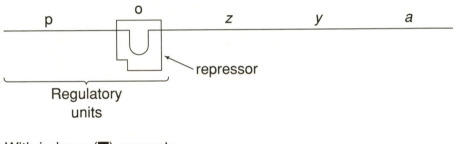

With inducer (■) present:

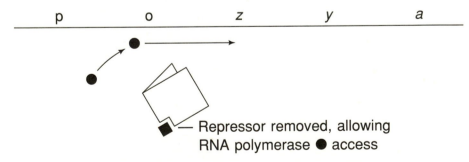

z, y, a sequences are transcribed, giving polycistronic RNA.
p = promoter, o = operator.

Example Metallothionine (Mt) gene expression is induced in the presence of heavy metals. Heavy-metal interactions with cellular receptors result in large increases in transcription of the Mt gene.

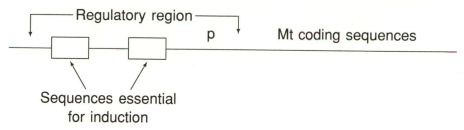

16 70 72

NOTE: In contrast to cis-acting regulatory units, **70 72** , inducers and repressors are **trans-acting factors,** that is, they can affect nonadjacent regions.

74

In eukaryotes, many genes are members of gene families.

A gene family is a group of genes with similar structure.

Example Tandem repeats of identical ribosomal RNA genes in nucleolus-organizer region of chromosome

| rRNA | rRNA | rRNA | rRNA |

These may have arisen by gene duplication due to unequal crossing over (at homologous-flanking sequences for example) **62** .

Example Gene families may contain nonfunctional units which differ slightly in structure from the active genes. These nonfunctional units are termed **pseudogenes.** A case is seen in subunit 9 of the *Neurospora* ATPase enzyme.

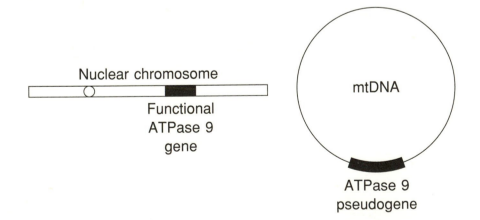

Nuclear chromosome

Functional
ATPase 9
gene

mtDNA

ATPase 9
pseudogene

Example   Mammalian globin genes (nonidentical)

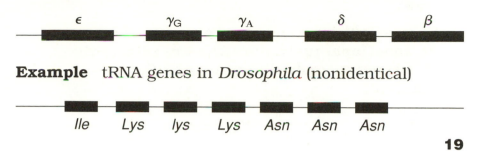

Example   tRNA genes in *Drosophila* (nonidentical)

19

NOTE: In gene families, each gene is expressed independently from its own promoter in contrast with prokaryotic operons **71** .

75

In most eukaryotic genes the sequences coding for functional protein or RNA are interrupted by sequences (called introns) of unknown function.

The existence of introns was a surprising fact arising out of the molecular analysis of gene structure, and was not predicted by classical genetic analysis.

Example A gene whose coding sequence is interrupted by one intron

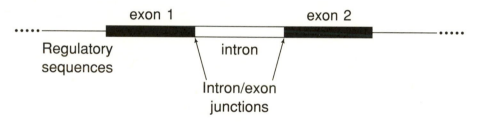

Many genes have multiple introns. They vary considerably in size and number from gene to gene, but the introns of any one gene are relatively invariant.

At the junction are found conserved consensus sequences.

5′ ‖‖ ▬▬▬▬▬▬▬   ▬▬▬▬ ‖‖ 3′

     AGGT   (Py)$_n$ AGG

$$(Py = pyrimidine).$$

These consensus sequences in HnRNA are probably recognized by **SNURP**s — small particles in the nucleus composed of RNA and protein that play a role in "processing."

One model

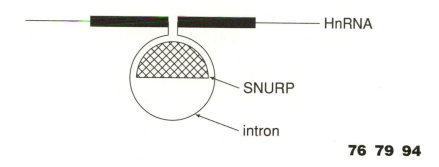

HnRNA

SNURP

intron

76 79 94

76

In eukaryotes the initial transcript of many structural genes is a precursor RNA (part of the HnRNA* population that is processed to mRNA).

The protein coding sequence is interrupted by sequences termed intervening sequence or *introns* which are excised from HnRNA to form mRNA **75** .

*HnRNA: Heterogeneous nuclear RNA, the mixed population of unprocessed transcripts found in the nucleus.

Example β globin gene

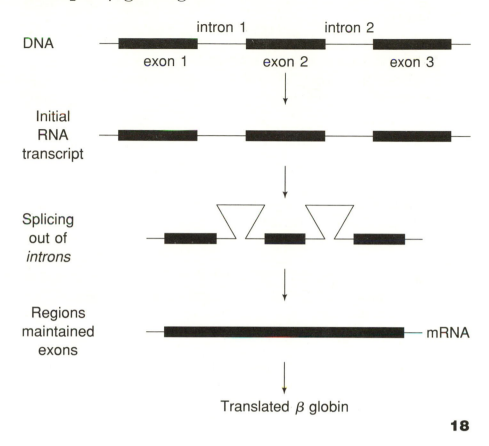

77

Eukaryotic genomes have a large proportion of their DNA in multiple copies.

DNA is often classified into three types:

1. *Unique DNA*—1 copy per *n* genome. About 30 to 75 percent of DNA. Contains most transcribed structural genes.

2. *Moderately repetitive DNA*—Up to 10^3 copies. From 1 to 30 percent of DNA, for example, histone and rRNA gene families **74** .

3. *Highly repetitive DNA*—Up to $\geq 10^6$ copies. From 5 to 45 percent of genomic DNA. Functions largely unknown, for example, in humans

   - *Eco RI DNA family*—2×10^5 copies each about 400 bp, arranged as tandem copies around the centromere.

   - *Alu I DNA family*—5×10^5 copies, each about 300 bp, dispersed with unique DNA.

Examples Autoradiograms of Southern blots **91** of restriction enzyme digested DNA **83** .

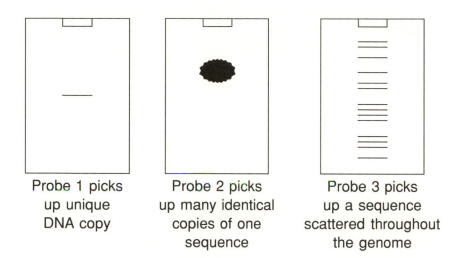

Probe 1 picks up unique DNA copy

Probe 2 picks up many identical copies of one sequence

Probe 3 picks up a sequence scattered throughout the genome

Prokaryotic and viral genomes have predominantly unique DNA, whereas eukaryote genomes have much higher proportions of repetitive DNA.

78

DNA contains several different kinds of sequences that are recognized and bound by various types of proteins.

Example   Origins of replication (ori)

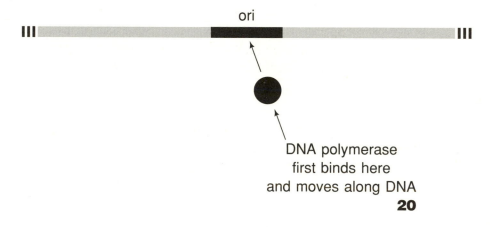

ori

DNA polymerase
first binds here
and moves along DNA

20

Example

- promotor (RNA polymerase binds)

- operator (repressor or inducer regulatory protein binds) **73** .

Example Restriction endonucleases

Such as Eco RI, which recognize a specific nucleotide sequence and cut the DNA usually within that sequence.

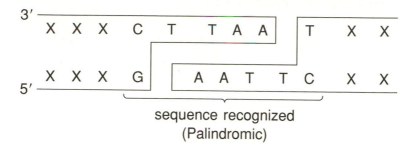

sequence recognized
(Palindromic)

NOTE: Staggered cut typical of many restriction enzymes.

81 82 83 108

157

79

RNA contains several different kinds of functional signal sequences.

Example   Prokaryotic mRNA

```
              1    2                    3
mRNA ─────────▬▬─▬▬────────────────▬▬─────────
```

1. Ribosome binding site (Shine-Delgarno sequence)

2. Translational start site (usually AUG)

3. Translational stop (usually UAG, UAA or UGA)

Example Eukaryotic HnRNA **76**

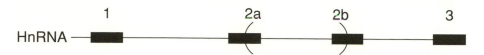

1. Cap site
2. Intron splice sites (a) acceptor; (b) donor
3. Polyadenylation site

Example Eukaryotic mRNA

4. Ribosome binding site
5. Translational start site (usually AUG)
6. Translational stop site (usually UAG, UAA or UGA)

80

The function of a structurally normal gene can be changed by its relocation to a new chromosomal domain.

Example A *Drosophila* w^+/w^- heterozygote in which w^+ is brought next to heterochromatin by an inversion.

64

w^+ is inactivated in some cells, causing position effect variegation of the eye; that is, a mixture of red (w^+) and white (w^-) areas.

Example One kind of human cancer, Burkitt's lymphoma

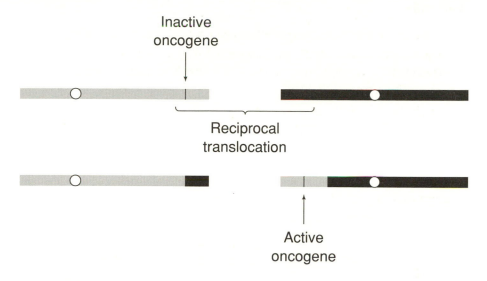

Inactive
oncogene

Reciprocal
translocation

Active
oncogene

Example X inactivation in female mammals

One of the X chromosomes and all its constituent genes is inactivated early in development **61** .

In X-autosome translocations, the inactivation can spread into the autosome.

81

Bacterial restriction enzymes will cut the DNA isolated from any organism into a set of defined fragments which differ for each restriction enzyme.

Restriction enzymes are the basic tools for a great deal of the techniques of gene manipulation.

RESTRICTION-ENZYME TARGET SITES 78

Example   Enzyme EcoRI.

$$5' \text{----} G | A \quad A \quad T \quad T \quad C \text{----} 3'$$
$$3' \text{----} C \quad T \quad T \quad A \quad A | G \text{----} 5'$$
staggered cut

Example   Enzyme Sma I

$$5' \text{----} C \quad C \quad C | G \quad G \quad G \text{----} 3'$$
$$3' \text{----} G \quad G \quad G | C \quad C \quad C \text{----} 5'$$
blunt end cut

The positions of the target sequences are constant features of the genome of an organism, for example, for a hypothetical organism with one simple linear chromosome (E = EcoRI cutting sites).

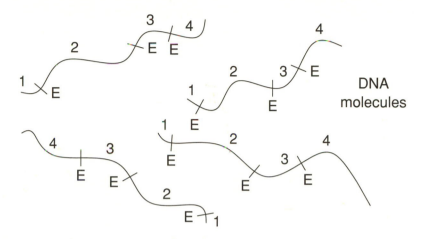

DNA molecules

Hence, samples of specific DNA fragments can be collected by electrophoresis.

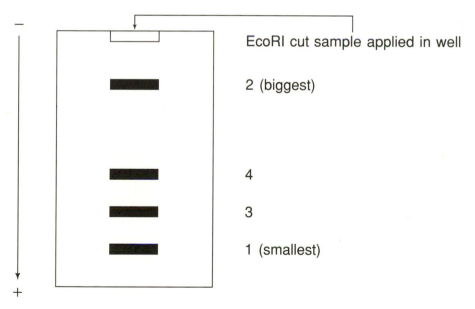

EcoRI cut sample applied in well

2 (biggest)

4

3

1 (smallest)

Double stranded DNA fragments are visualized by staining with ethidium bromide.   **82 83 85 86 88**

82

Restriction-enzyme target sites can be mapped on a DNA molecule.

Example Mapping a fragment of DNA using two different restriction enzymes

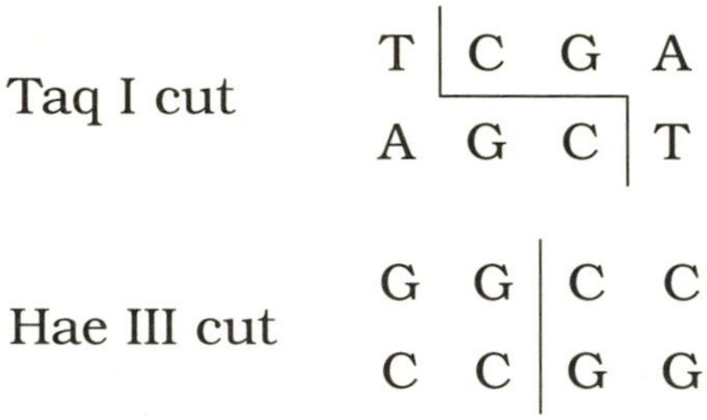

Taq I cut

$$\begin{array}{c|ccc} T & C & G & A \\ \hline A & G & C & T \end{array}$$

Hae III cut

$$\begin{array}{cc|cc} G & G & C & C \\ C & C & G & G \end{array}$$

78 81 83

Cut samples of a linear DNA molecule with Taq I and Hae III separately and together. Fragment lengths correspond to the distance between target cut sites on the DNA.

samples

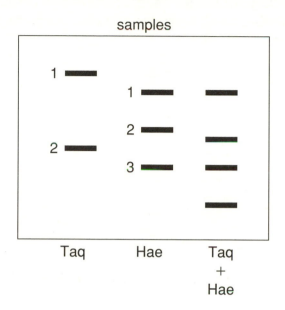

Rough map must be

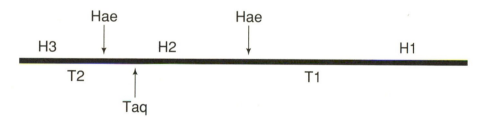

83

Restriction-enzyme target sites are convenient landmarks for the study of DNA structure and function.

1. They define specific segments of DNA for study.

2. They act as markers to which genes can be mapped.

Example From a restriction map

gene *x* (Only detectable in
this fragment so its locus is
mapped roughly)

Example Restriction fragment length polymorphisms (RFLPs) act as heterozygous chromosomal markers.

Hind III(H)

```
H    III   H   II   H          I        H
├──────────┼─────┼────────────────────┤ DNA haplotype* A¹
├──────────────────┼────────────────────┤ DNA haplotype A²
H        IV        H         I         H
```

An individual (say) $A^1A^2\ leu^+\ leu^-$ is a dihybrid suitable for studying recombination.

If the products of meiosis of such a dihybrid were (say)

$$40\%\ A^1\ leu^+$$
$$40\%\ A^2\ leu^-$$
$$10\%\ A^1\ leu^-$$
$$10\%\ A^2\ leu^+$$

We could infer linkage as follows:

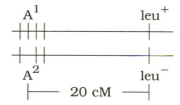

*Haplotypes ("genotypes") defined by Hind III fragments are:

$$A1 = I + II + III$$
$$A2 = I + IV$$

1 78 81 82 108

167

84

Single stranded nucleic acids with complementary stretches of nucleotide sequence will find and hybridize to each other in solution.

This homology at the level of nucleotide sequence is another of the fundamental processes on which genome manipulation is based.

AT THE LEVEL OF BASES

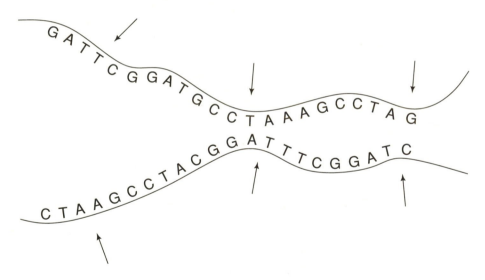

Hybridization is by trial and error in random collisions.

TWO BASIC TYPES OF HYBRIDIZATION

1. ssDNA with ssDNA

2. ssDNA with RNA

17 18

Example In situ hybridization

- Partially denature *Drosophila* polytene chromosome.
- Bathe in a specific radioactive tRNA probe **92** .
- Complementary sequences bind, others wash off.
- Expose to X-ray film (which is sensitive to radioactivity).
- Autoradiograph.

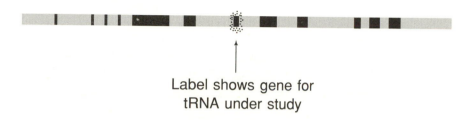

Label shows gene for
tRNA under study

Example The course of hybridization of single-stranded DNA in solutions can be followed by means of Cot curves.

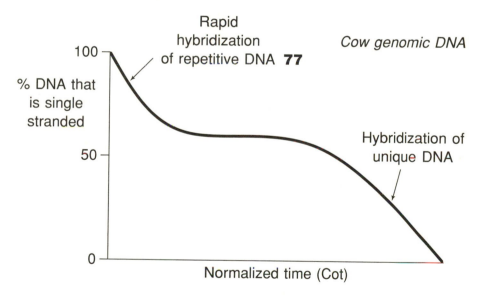

85

DNA fragments from diverse sources can be joined together in vitro, using ligase enzyme, to form a recombinant DNA molecule.

This principle provides the basis of recombinant DNA technology, which allows the creation of completely novel genomes by the combination of unrelated DNA molecules. Note the difference from homologous recombination **22** .

The fragments to be joined may have sticky ends (e.g., cut with EcoRI) or blunt ends (e.g., cut with Sma I **81 86**)

Example Sticky-end ligation

- Mouse DNA fragment generated by EcoRI

```
AATTC ————————————————————— G
    G ————————————————————— CTTAA
```

- Bacterial plasmid cut by EcoRI

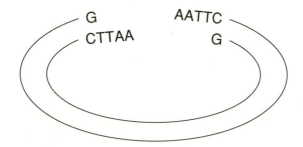

```
    G          AATTC
    CTTAA          G
```

- Recombinant DNA molecule

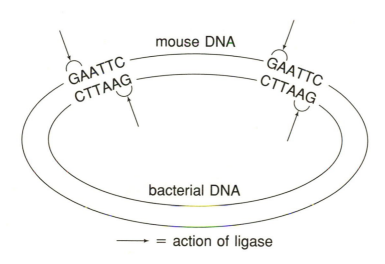

mouse DNA

GAATTC
CTTAAG

GAATTC
CTTAAG

bacterial DNA

⟶ = action of ligase

86

A DNA fragment can be amplified for study by inserting it into a DNA-cloning vector ("carrier") and propagating in a rapidly growing host.

This principle provides a way of obtaining large amounts of specific DNA regions for study.

Example The specially designed *E. coli* R plasmid **29** pBR322 has several cloning sites, one of which is a Bam HI site within a tetracycline-resistance gene. Plasmids with inserts confer amp resistance on bacteria but the tet resistance is lost due to the insertion.

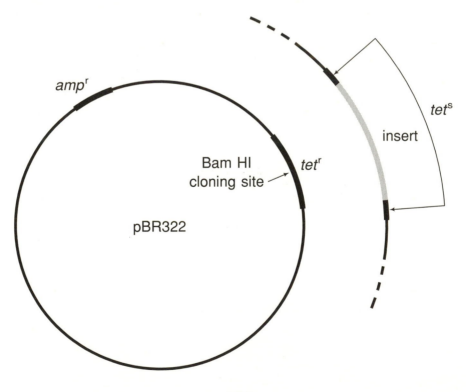

Example λ phage as a vector **31**

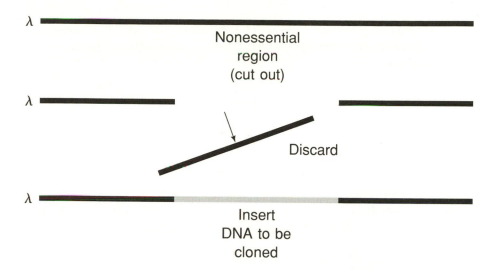

Insert DNA fragment is needed to produce correct size for packing into phage head. **78 81**

87

New DNA can be introduced into cells by transformation, transfection, microinjection, or on microprojectiles.

It is perhaps surprising that extrinsic DNA is taken up by both prokaryotic and eukaryotic cells, and that it can be maintained in cells in a stable manner, either replicating autonomously or integrated into the genome.

Example Transforming $amp^s tet^s$ *E. coli* with recombinant pBR322 plasmid DNA **29 86** .

*   Plasmid DNA

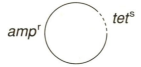

amp^r tet^s

*   Recipient bacteria cell

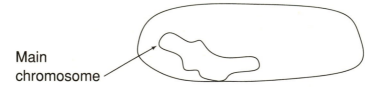

Main chromosome

*   Add plasmid DNA to $CaCl_2$-treated cells
*   Select $amp^r tet^s$ transformants

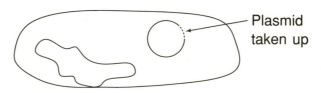

Plasmid taken up

Example   Transfection of *E. coli* with recombinant λ-phage DNA results in phage progeny (plaques) **86** .

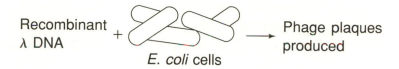

Recombinant λ DNA + E. coli cells → Phage plaques produced

Example   Correcting growth deficiency in dwarf mice by adding rat growth-hormone (GH) gene spliced to a mouse heavy-metal-sensitive promotor **73 94** .

- Inject fertilized dwarf egg with recombinant DNA.

- Implant into female.

- Some progeny are of normal size and have inserted chimeric DNA in a mouse chromosome.

Example   Plant and fungal cells have been successfully modified by firing into them tungsten microprojectiles coated with the appropriate DNA.

Plants or animals whose genome has been modified by the insertion of externally-added DNA are called *transgenic* organisms.

88

An entire genome can be cloned as fragments in a heterogeneous population of microbial cells or viruses that is called a genomic library.

The purpose of the library is to maintain the genome as conveniently sized fragments, any of which can be "withdrawn" from the library for study.

Example Eukaryotic genomic library in λ phage

1. Isolate eukaryotic DNA.

2. Cut with a restriction enzyme such as EcoRI **78** .

3. Ligate in λ strain Charon 4a **86** .

4. Transfect *E. coli* and obtain plaques. Only λ with inserts will form plaques **87** .

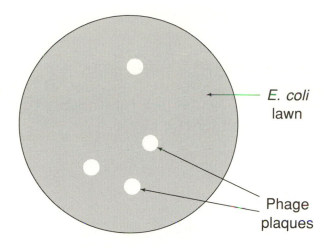

E. coli lawn

Phage plaques

31

- Each plaque represents a cloned eukaryotic fragment.

- How many plaques represent an entire genomic library?

- Assume eukaryotic genome is 10^{10} nucleotide pairs; Charon 4a(λ) is known to carry 1.5×10^4 nucleotide pairs, therefore, would need at least $10^{10} \div (1.5 \times 10^4) \approx 6 \times 10^5$ plaques.

See also: cDNA libraries **90**

89

Specific clones can be selected from a genomic library by using a hybridizing probe of complementary sequence 84 or by detecting function of a gene carried by the clone.

Example *Probing* the colonies that constitute the library **92** . (Using radioactively labelled specific DNA sequence)

Cells lysed, DNA denatured and ssDNA transferred to cellulose nitrate paper

Cellulose nitrate bathed in probe

Radioactive spot due to the radioactive probe specifically hybridizing to the DNA of this colony

Plaque, or colony, hybridization

Example *Complementing* auxotrophic recipient cells with cloned DNA

Plasmid-based library
made from *leu*$^+$ genome
of yeast **86 88**

↓

Transform *leu*$^-$ recipient
yeast cells **87**

Plate cells on
leucineless medium

Contain
leu$^+$-bearing
plasmid **50**

90

Protein-coding regions of eukaryotic genes can be obtained by synthetizing cDNA* from mRNA populations.

Thus, we see there are two basic kinds of libraries, genomic libraries made from fragmented DNA and **cDNA libraries** made from mature mRNA.

Example   Making a cDNA library.

1. Extract mRNA population (extraction based on the fact that mRNA tails are polyadenylated.)

*Complementary DNA

2. Make cDNAs from mRNAs using reverse transcriptase enzyme.

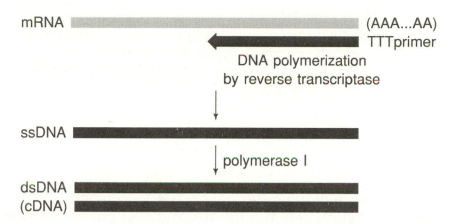

3. Poly C tails added to the 3′ ends of the c DNA and poly G tails to the 3′ ends of the vector DNA to permit ligation   **86**

4. Ligate into plasmid vector creating cDNA library in bacteria   **88**

5. Screen library as in **89**

91

A specific nucleic acid fragment can be detected amongst a heterogeneous population of fragments through hybridization with a probe, a labelled nucleic acid with complementary nucleotide sequence.

Here we see a combination of restriction enzyme, electrophoretic, hybridization, and autoradiographic techniques uniting to provide a powerful analytical tool.

Example Southern blotting

1. Electrophorese DNA fragments

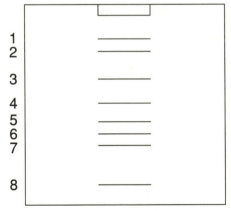

Bands stained with
ethidium bromide

18 78 81

182

2. Blot fragments into nitrocellulose

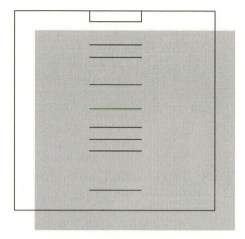

3. Bathe nitrocellulose in probe, for example, specific radioactive tRNA

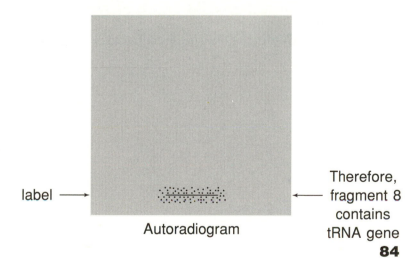

label ⟶

Autoradiogram

⟵ Therefore, fragment 8 contains tRNA gene

84

92

A hybridization probe 91 can be a specific RNA, a specific DNA or cDNA, or a specific synthetic oligonucleotide.

Success in molecular cloning is dependant on the availability of a suitable probe to detect the DNA of interest.

1. **RNA.** For example, tRNA, rRNA.

2. **cDNA.** Some tissues are enriched for a specific mRNA, for example, in chicken oviduct cells, 40 percent mRNA is ovalbumin mRNA. Many cDNA clones from such tissue will be ovalbumin cDNA.   **90**

3. **DNA.** Equivalent DNA cloned from a related organism.
      97

4. **Oligonucleotides.** If at least part of a protein amino acid sequence is known, a nucleotide sequence can be deduced and a short DNA molecule with this sequence can be made chemically. Hence the gene for that protein can be selected from a genomic library.   **88**

NOTE: Because of redundancy of the genetic code, a *mixture* of synonymous oligonucleotides is used as a hybridization probe.

93

The nucleotide sequence of a cloned DNA fragment can be obtained using the Maxam-Gilbert or the Sanger techniques.

Perhaps the most fundamental property of a gene is its nucleotide sequence, and this can now be deduced using a variety of techniques, some of which have been automated. Some smaller entire genomes have been fully sequenced.

Example   Maxam-Gilbert

1. Cut DNA out of vector, denature, and end-label.

2. Chemically treat to break *rarely* at G, C, T, or A.

3. Electrophorese and autoradiograph.

G   A   C   T

4. Sequence is read from base of gel upwards:

C C T T G A T A G C A G T

Structurally important regions to study

- 5′ regulatory regions **72**
- Reading frames
- Intron – exon junctions **75**

94

Expression of a gene can be engineered (for example to the specific organism or to some specific environmental stimulus) by making a chimeric gene consisting of the transcribed region ligated to an appropriate control region

Here we see that recombinant-DNA technology allows the creation of DNA constructs tailored for some specific experimental or commercial purpose.

Example The use of the *E. coli lac Z* gene as a "reporter gene" to study the developmental specificity of Drosophila heat shock promoters:

Chimeric gene

Heat shock promoter	lac Z gene
(*Drosophila*)	(*E. coli*)

Transgenic **87** *Drosophila* bearing this construct are heat shocked, then killed and stained for *Z* activity; such tissues turn blue because of a blue dye formed from a marker substrate by β-galactosidase (*Z*) activity.

Example Expression of rat growth hormone (GH) transcript occurs in the presence of heavy metals (see also **73**).

Chimeric gene

| Mouse Metallothionine gene promoter region (sensitive to heavy metal) | Rat GH sequence |

85

Example Production of mammalian insulin in bacteria

Chimeric gene

| Bacterial promotor region | Intron-free* insulin coding sequence |

*Introns cannot be processed in bacteria.

75

95

**Any gene with an identifiable mutant pheno-
type can in principle be isolated by transposon
mutagenesis, in which transposons are used to
cause mutations and thereby act as molecular
tags for the desired piece of DNA.**

Hence, if the transposon is rare in the genome, probing
for the transposon should lead to the isolation of the
mutant gene.

PROCEDURE

1. Select mutatant phenotype for gene *x* in a strain into
   which active transposons have been introduced.

   Some mutations will be thus:

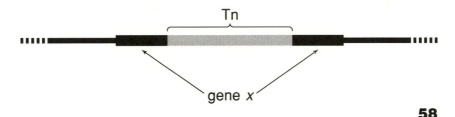

58

2. Use a cloned transposon as a probe

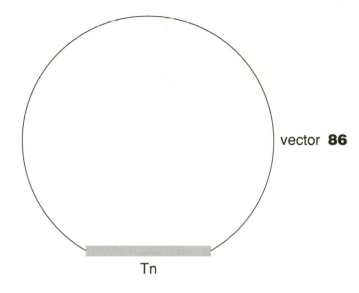

vector **86**

Tn

3. This can be used to recover interrupted gene *x* from a library by hybridization

89

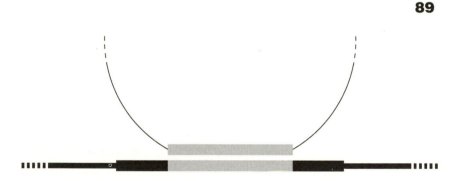

4. Use subcloned end fragment to recover complete gene *x* from wild-type library.

5. Test for *x* by transformation **87** or by in situ hybridization **84** .

96

Regions of the genome that cannot be cloned directly can be selected from a library by chromosome-walking from an adjacent cloned region

This technique relies on the availability of a closely linked gene that is already cloned or is cloneable.

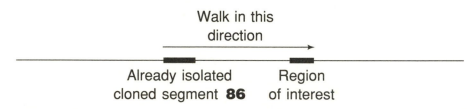

Walk in this direction

Already isolated cloned segment **86**   Region of interest

PROCEDURE

1. Take a subclone from present cloned segment.

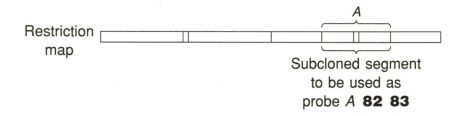

Restriction map

A

Subcloned segment to be used as probe *A* **82 83**

2. New fragment recovered from library using probe *A*

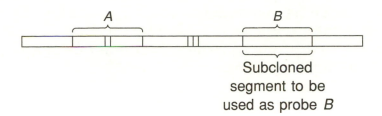

Subcloned
segment to be
used as probe *B*

3. Repeat until overlaps reach region of interest.

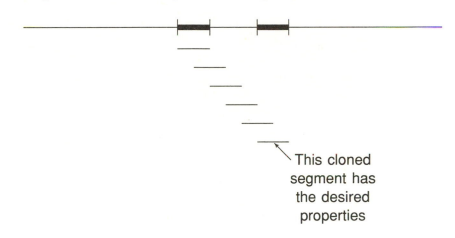

This cloned
segment has
the desired
properties

97

If a gene has been evolutionarily conserved, a clone of such a gene from one species can be used as a probe to isolate homologous genes from other species.

Many essential genes are conserved, for example, cytochrome c, globins

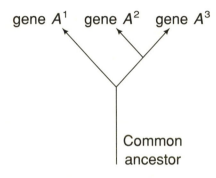

Common ancestor

1, 2, and 3 show strong sequence homology and would hybridize to each other **84 91** .

Example Restriction map of mitochondrial DNA from relatively unstudied fungus "P" **1 82**

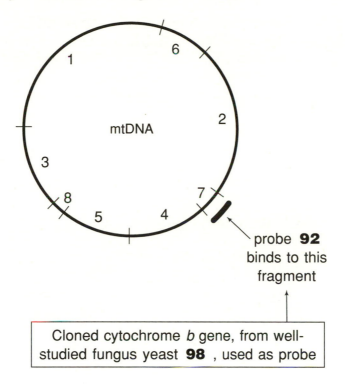

probe **92** binds to this fragment

Cloned cytochrome *b* gene, from well-studied fungus yeast **98** , used as probe

98

Mitochondria and chloroplasts have their own unique circular DNA "chromosomes" distinct from nuclear DNA.

The genes of these organelles are mainly concerned with energy production or organelle-specific protein synthesis.

Yeast mt DNA, with some genes shown

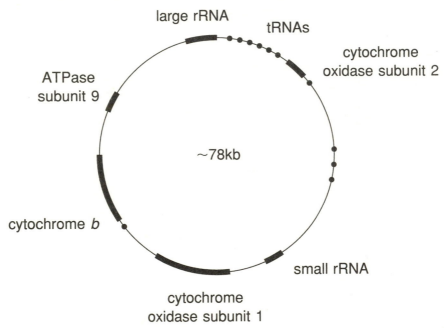

1

196

Spinach cp DNA, with some genes shown

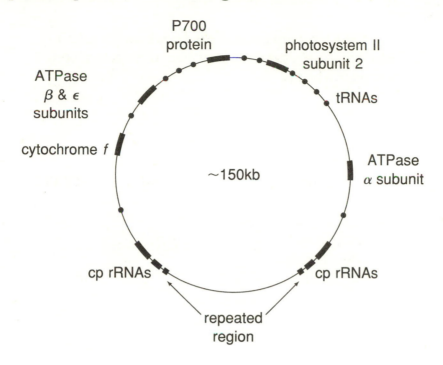

99

mt DNA and cpDNA, and their associated phenotypic variants are usually inherited uniparentally.

AN OPERATIONAL TEST OF ORGANELLE MUTATIONS

Example Variegated higher plant

Inter-sector crosses:

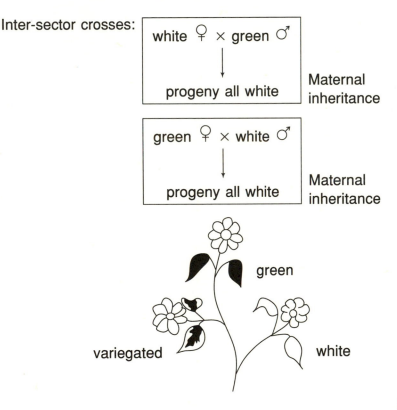

white ♀ × green ♂

↓

progeny all white

Maternal inheritance

green ♀ × white ♂

↓

progeny all white

Maternal inheritance

green

variegated

white

98

Example Some yeast antibiotic-resistance mutants

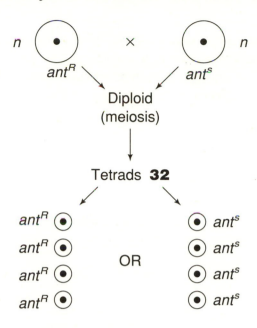

Example Senescent *Neurospora*

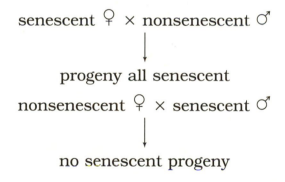

100

In cytoplasms consisting of a mixture of organellar variants, a sorting-out process occurs by an unknown mechanism, so that many daughter cells are pure for one variant type.

This sorting-out process is not understood, but is observed in several different organisms.

Example Yeast

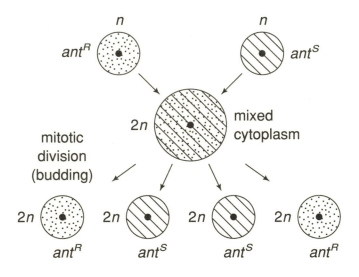

Example Higher plants

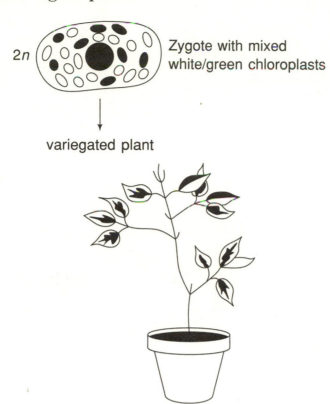

$2n$ Zygote with mixed white/green chloroplasts

variegated plant

98 99

201

101

The variant phenotypes associated with organellar DNA variants are transmitted by cytoplasmic contact alone.

Example The heterokaryon test

Fungal strain 1

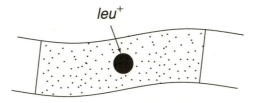

leu$^+$

Senescent mitochondrial variant

Fungal strain 2

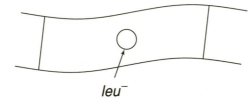

leu$^-$

Normal mitochondrial type

Heterokaryon

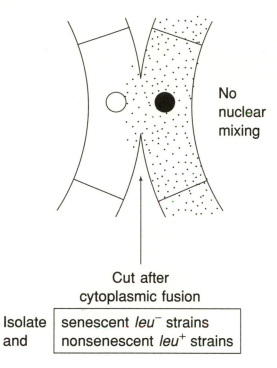

No
nuclear
mixing

Cut after
cytoplasmic fusion

Isolate and	senescent *leu⁻* strains
	nonsenescent *leu⁺* strains

NOTE: This is another operational criterion for organelle-based mutation **98 99 100** .

203

102

Several organelle-located proteins have subunits coded in both the nuclear and the organellar DNA.

Here we see the molecular basis of many nucleocytoplasmic genetic interactions.

Example Yeast mitochondrial membrane ATPase

- Seven nuclear DNA subunits (genes)

- Three mtDNA subunits (genes)

Only four of the seven DNA subunits are shown in the drawing at right.

Example Photosynthetic proteins in the chloroplast membranes **98**

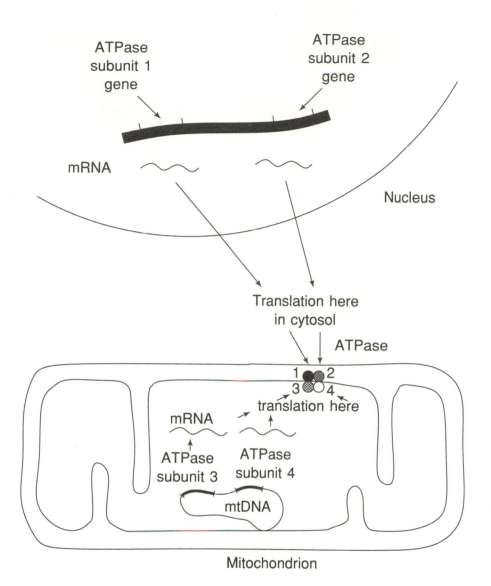

ATPase subunit 1 gene

ATPase subunit 2 gene

mRNA

Nucleus

Translation here in cytosol

ATPase

1 ● ● 2
3 ● ○ 4
translation here

mRNA

ATPase subunit 3

ATPase subunit 4

mtDNA

Mitochondrion

103

In several well-studied organisms, organelle DNA is spontaneously rearranged at high frequency.

In the plant example there is no associated mutant phenotype; the processes appear to be tolerated.

Example   *Plant mtDNA rearrangement*

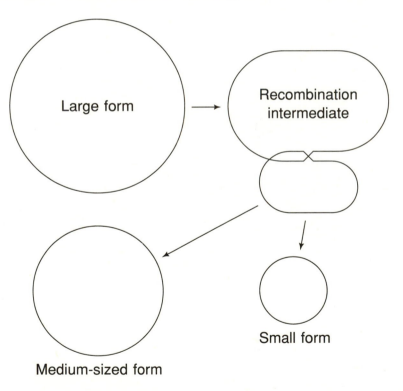

98

Example   *Petite yeast*

Normal yeast mtDNA

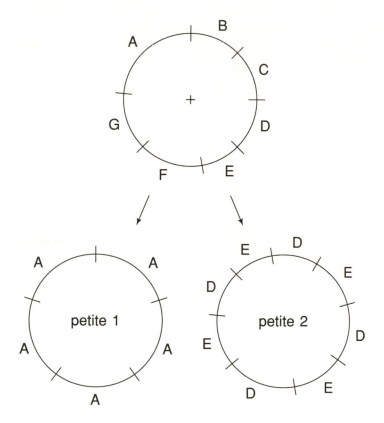

Petites thus represent deletions and as such can be used in mapping mtDNA **60** .

104

Phenotypic variance in a population is made up of two components, genetic variance and environmental variance, whose relative sizes are measured by broad heritability (H^2).

This principle concerns the problem of how to disentangle genetic and environmental contributions to quantitative variation.

Phenotypic continuous variation **1** is measured by variance

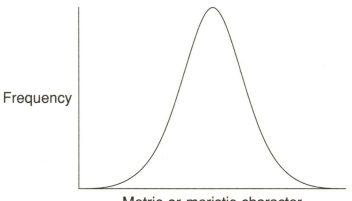

The components of variance (S^2) are additive.

Hence,

$$S^2_{pheno} = S^2_{geno} + S^2_{enviro}$$

Broad sense heritability $= \boxed{H^2 = \dfrac{S^2_{geno}}{S^2_{pheno}}}$

105 106

EXPERIMENTAL ESTIMATIONS

1. Obtain asexual clones of several individual genotypes.

Clone of one genotype, obtained from cuttings

All variance within a clone should be environmental, so estimate S^2_{geno} in the population as

$$S^2_{pheno} \quad minus \quad S^2_{enviro}$$
(population variance)   (mean intraclonal variance)

2. Create many individuals of identical genotypes, for example,

$$AAbbCCDD \times aaBBccDD$$

Identical progeny $AaBbCcDD$

Variance in progeny must be environmental.

H^2 is calculated in a manner similar to that used in item 1 above.

105

The proportion of phenotypic variance that responds to artificial selection is called the additive genetic variance, and its relative size is measured by the narrow heritability, h^2.

There are two major genetic effects that contribute to genetic variation: various degrees of dominance, and the additive effects of genes. The additive effects are most useful as a basis for reliable selection.

$$S^2_{pheno} = \underbrace{S^2_{additive} + S^2_{dominance}}_{S^2_{geno}} + S^2_{enviro}$$

$$\boxed{h^2 = \frac{S^2_{additive}}{S^2_{pheno}}}$$

1 104 106

EXPERIMENTAL ESTIMATIONS

1. By selection experiments

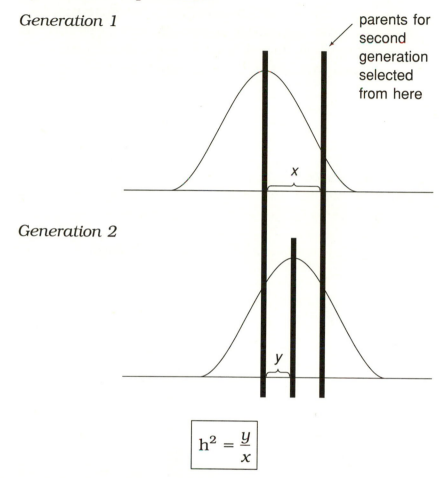

Generation 1

parents for
second
generation
selected
from here

x

Generation 2

y

$$h^2 = \frac{y}{x}$$

2. By measuring correlations between relatives

Example $h^2 = 2$ (Correlation coefficient of phenotypes of siblings)

106

The genetic variation involved in continuous phenotypic variation is attributed to variation in polygenes, genes that have similar and cumulative effects on a phenotype.

Example Three hypothetical loci affect one phenotype and each capital letter allele represents one dose of "activity."

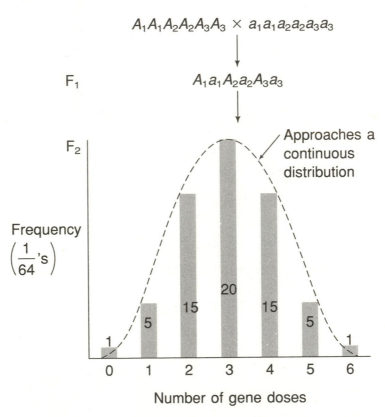

$$A_1A_1A_2A_2A_3A_3 \times a_1a_1a_2a_2a_3a_3$$

F_1

$$A_1a_1A_2a_2A_3a_3$$

F_2

Approaches a continuous distribution

Frequency $\left(\frac{1}{64}\text{'s}\right)$

20

15   15

5   5

1   1

0   1   2   3   4   5   6

Number of gene doses

Typical experimental segregation patterns attributed to polygenes

Parents

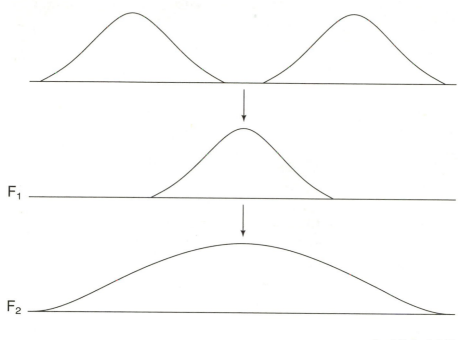

F₁

F₂

1 104 105

107

Some discontinuous variation is caused by polygenes with a threshold effect of gene dose.

Examples Some probable examples of threshold effects in humans are:

- Cleft palate or lip
- Club foot

Pedigrees of such phenotypes show recurrence from generation to generation but no simple patterns.

MECHANISM

For example, three hypothetical polygene loci contribute to a certain character so that:

- If an individual has five or more gene doses, phenotype A is shown.

- If an individual has four or fewer gene doses, phenotype B is shown.

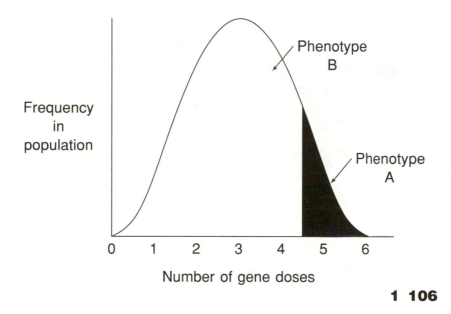

1 106

215

108

Populations show a surprisingly high level of discontinuous variation attributable to single Mendelian loci.

MORPHOLOGICAL POLYMORPHISM 1

Example Most populations of white clover are polymorphic for leaf chevrons

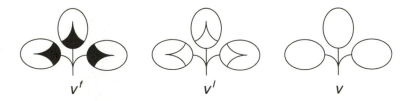

v^f v^I v

ALLOZYME POLYMORPHISM

Example Electrophoresis of enzyme x

1	—	—		—			
2			—	—	—		
3	—	—				—	—

Allozyme x^1x^3 x^3x^3 x^1x^1 x^2x^2 x^1x^2 x^2x^3 x^3x^3

In vertebrates, on average an individual is heterozygous for about 5 percent of enzyme loci, and populations are polymorphic for about 20 percent of enzyme loci with no apparent phenotypic differences.

DNA RESTRICTION-FRAGMENT-LENGTH POLYMORPHISMS (RFLP)

Two basic types:

1. Associated with change in gene function, for example, thalassemia—a disease caused by mutant β-globin gene: restriction site changed by mutation

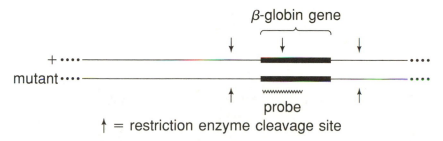

β-globin gene

probe

↑ = restriction enzyme cleavage site

2. Those associated with no apparent phenotypic differences, for example, two individuals might have identical functional β-globin genes but a difference in a restriction site outside the gene.

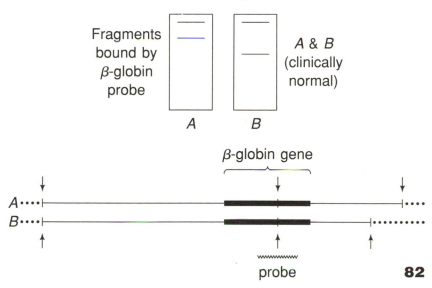

Fragments bound by β-globin probe

A   B

A & B (clinically normal)

β-globin gene

probe   **82**

Both *A* and *B* are homozygous in the cases shown on the gel.

RARE MUTATIONS 39 44.

217

109

In populations showing discontinuous variation attributable to a single Mendelian locus, a basic description of the genetic structure of the population is measurement of allele frequencies.

Example   For an allozyme locus A with two alleles 1 and 2

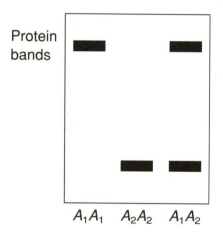

Protein bands

A_1A_1 A_2A_2 A_1A_2

Let

Homo (1) = Proportion of A_1A_1 individuals

Homo (2) = Proportion of A_2A_2 individuals

Het = Proportion of A_1A_2 individuals

Note that:

Homo (1) + Homo (2) + Het = 1.

The frequency of one allele (here A_1) is p.

$$p = \text{Homo (1)} + \tfrac{1}{2}\text{Het}$$

The frequency of the other allele (here A_2) is q.

$$q = \text{Homo (2)} + \tfrac{1}{2}\text{Het}$$

Note that $p + q = 1$.

For more alleles, say 3, an extra term is introduced, for example, $p + q + r = 1$.   **1 108**

110

In a large interbreeding diploid population with constant allele frequencies, the genotypic proportions for an autosomal locus with two alleles are stable and given by the Hardy-Weinberg formula.

Other genotypic proportions are possible for any specific p and q value, but by definition those are not equilibrium proportions. This is how the Hardy-Weinberg equilibrium works:

HARDY-WEINBERG EQUILIBRIUM FORMULA

p^2	$2pq$	q^2
A_1A_1	A_1A_2	A_2A_2

NOTE: $p^2 + 2pq + q^2 = 1$.

109 111

Note the equilibrium:

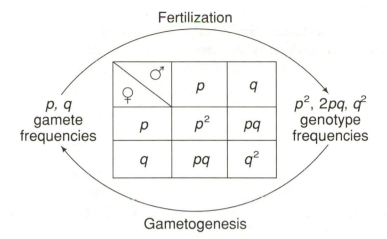

If Hardy-Weinberg equilibrium is assumed, the frequencies of genotypes underlying a dominant phenotype can be calculated, for example, for albinism (aa), the frequency of affected individuals = $1/10,000 = q^2$. Hence:

$$q = \sqrt{\frac{1}{10000}} = \frac{1}{100}$$

Therefore

$$p = 1 - \frac{1}{100} = \frac{99}{100}.$$

Then p^2 and $2pq$ can be calculated: $2pq$ (the frequency of heterozygotes) is

$$2 \times \frac{99}{100} \times \frac{1}{100} \approx \frac{1}{50}$$

111

For populations in Hardy-Weinberg equilibrium the proportion of heterozygotes is maximal when allele frequencies are equal.

The calculation of Hardy-Weinberg genotypic proportions for different allele frequencies reveals the basis of this principle, together with other important insights into population structure.

Example For a gene with two alleles A and a, the frequency of heterozygotes is highest when $p = q = 1/2$.

109 110

This graph shows all possibilities.

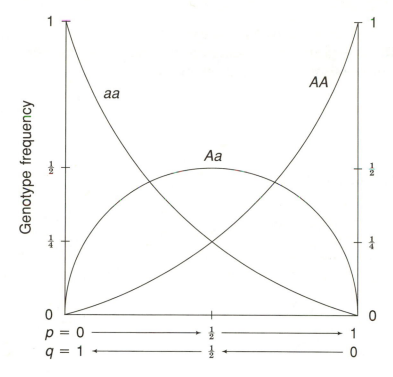

NOTE: Most rare alleles are in heterozygotes (▨).

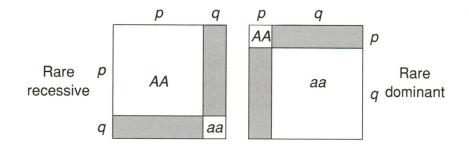

112

For populations in Hardy-Weinberg equilibrium, the frequencies of all types of matings in the population can be calculated.

CALCULATION

♂ Parents / ♀ Parents	p^2 AA	$2pq$ Aa	q^2 aa
p^2 AA	p^4	$2p^3q$	p^2q^2
$2pq$ Aa	$2p^3q$	$4p^2q^2$	$2pq^3$
q^2 aa	p^2q^2	$2pq^3$	q^4

The grid shows the frequencies of the different matings.

109 110

NOTE: The frequency of matings of genotype Aa × genotype aa (for example) is $2pq^3 + 2pq^3 = 4pq^3$, but the frequency of matings of Aa females × aa males is $2pq^3$.

RELEVANCE

The breeding structure of the population is important in ecological and evolutionary studies.

The grid is also useful in genetic counseling of parents about the incidence of genetic disease.

113

Selection against a recessive phenotype is less efficient at low values of _q_.

This principle shows how selection works at the level of allele frequencies. In our example, we will use the most severe selection possible to illustrate the process.

Example Starting population

- $p = 1/2$, $q = 1/2$ **109**
- Impose complete selection against the _aa_ phenotype.

			A	a
p^2	_AA_	1/4	1/4	0
$2pq$	_Aa_	1/2	1/4	1/4
q^2	_aa_	1/4	0	~~1/4~~

Gametes

$$p' = 2/3, q' = 1/3$$

q will go $1/2 \rightarrow 1/3 \rightarrow 1/4 \rightarrow 1/5 \rightarrow 1/6$ etc.

$$q \qquad q' \qquad q'' \qquad q''' \qquad q''''$$

over successive generations. **110 111**

Here is a graphic representation of the data:

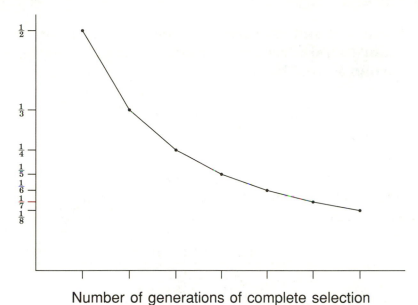

Number of generations of complete selection

Hence, for a rare recessive allele to go from $q = 1/100 \rightarrow$ $q = 1/200$ would take 100 generations of selection.

114

Recurrent net mutation in one direction can change allele frequency until a mutation equilibrium is reached.

We will consider two alleles A_1 and A_2, that are interconverted at different rates. In this example we will ignore selection.

$$A_1 \xrightarrow{\text{Mutation rate } \mu} A_2$$
$$\xleftarrow{\text{Mutation rate } \nu}$$

If μ is forward mutation, μ is usually $> \nu$, and there is net mutation pressure from A_1 to A_2.   **39 44 109 115**

Mutation equilibrium is reached when frequency of A_2 is

$$\hat{q} = \frac{\mu}{\mu + \nu}$$

SIMULATION

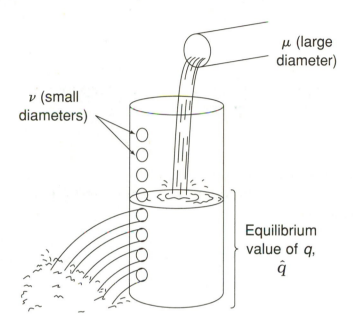

μ (large diameter)

ν (small diameters)

Equilibrium value of q, $\hat{q}$

NOTE: Mutation pressure is very slow acting.

115

When selection and mutation act in opposite directions, stable allele frequencies are reached, called a selection/mutation equilibrium.

This principle shows the complex interplay of forces that determine allele frequencies.

Example

- If selection is against *aa* such that the fitness of *aa* is less than wild type by a proportion s (the selection coefficient).

- If mutation A to a is at a rate μ, and ν can be ignored:
  109 113 114

Selection/mutation equilibrium occurs when

$$\hat{q} = \sqrt{\frac{\mu}{s}}$$

SIMULATION

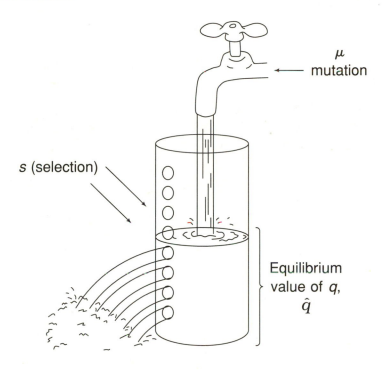

μ
mutation

s (selection)

Equilibrium
value of q,
$\hat{q}$

116

In any population of finite size, allele frequencies from generation to generation can change erratically by random genetic drift.

THE REASON

- *Sampling error.* Because of finite number, the gametes may not be a typical sample of the allele frequencies of the previous generation. **109 117**

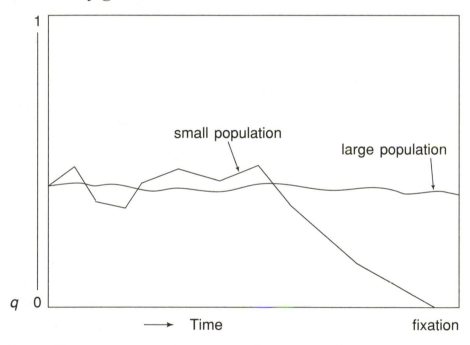

Generation 1

75% : 25%

Generation 2

80% : 20%

Over many generations:

small population

large population

q 0

Time   fixation

NOTE: Random genetic drift of allele frequencies is more pronounced the smaller the population.

233

117

Inbreeding influences the genotypic structure of a population by proportionately increasing homozygotes and decreasing heterozygotes.

Example Selfing, an extreme form of inbreeding.

Assume an initial population consisting entirely of heterozygotes

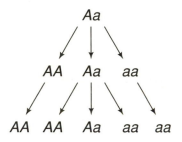

11 109 110

234

For specific values of p and q, and specific values of the inbreeding coefficient, at equilibrium the population structure is:

$$AA = pF + p^2(1 - F)$$
$$Aa = 2pq(1 - F)*$$
$$aa = qF + q^2(1 - F)$$

Where F is the inbreeding coefficient, a fraction that measures the average degree of inbreeding in a population.

Small populations show inbreeding effects even though no inbreeding is deliberately practiced.

*NOTE: This formula shows that F can also be thought of as the proportionate reduction of heterozygotes through inbreeding.

118

In a population, linked or unlinked loci will attain linkage equilibrium, in which genotypic ratios for each locus are combined randomly. Linkage merely slows the attainment of this equilibrium.

This principle shows the effects of linkage at the population level. The surprising conclusion is that these effects (in maintaining possible formable allele combinations) are merely temporary.

Example Linkage equilibrium of the A/a and B/b loci

B/b \ A/a	p^2 AA	$2pq$ Aa	q^2 aa
P^2 BB	AA BB	Aa BB	aa BB
$2PQ$ Bb	AA Bb	Aa Bb	aa Bb
Q^2 bb	Aa bb	Aa bb	aa bb

NOTE: Genotypic frequencies will be the product of frequencies for individual loci (e.g. p^2P^2 AABB)

26 109 110 111

236

At the gamete level, let gametes

$$AB = r$$
$$aB = s$$
$$Ab = t$$
$$ab = u$$

Linkage equilibrium occurs when

$$ru = st$$

Linkage disequilibrium is represented as

$$d = ru - st$$

Example Start with equal *AA BB* and *aa bb*

$r = \frac{1}{2}$, $u = \frac{1}{2}$, $s = 0$, $t = 0$
$d = \frac{1}{4} - 0 = \frac{1}{4}$
at equilibrium r, u, s, t, all $\frac{1}{4}$
$d = \frac{1}{16} - \frac{1}{16} = 0$

NOTE: $d_t = (1 - RF)d_{t-1}$ (RF = recombinant frequency)

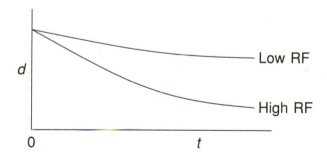

Hence, linkage is seen merely to slow attainment of linkage equilibrium ($d = 0$)

119

In populations, sex-linked loci show different equilibrium genotypic values for each sex.

In the XY system, most of the X chromosomes in the population belong to the female sex, so the equilibrium formulas have to be adjusted accordingly.

Example X-linked gene A/a

At equilibrium:

$$\female \begin{cases} AA\ p^2 \\ Aa\ 2pq \\ aa\ q^2 \end{cases}$$

$$\male \begin{cases} A(Y)\ p \\ a(Y)\ q \end{cases}$$

13 109 110

NOTE: Since p and q are always less than 1, this means that males will always show a higher proportion of X-linked recessive conditions.

Example where $q = 0.4$

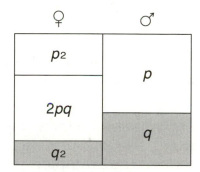

If not in equilibrium

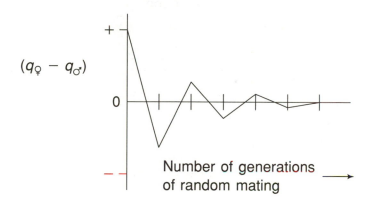

120

Reverse genetics provides a way of isolating a gene starting with its encoded protein.

- Both *traditional* and *reverse* genetics provide powerful ways of revealing how genes control biological functions.

- The *traditional* approach of genetics is to begin with a mutant phenotype, and end up with the appropriate protein malfunction. For example,

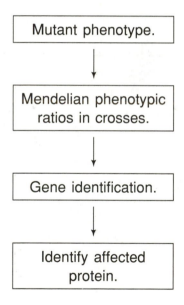

- *Reverse* genetics, as its name suggests, works the other way, using protein to track down the gene that encodes it.

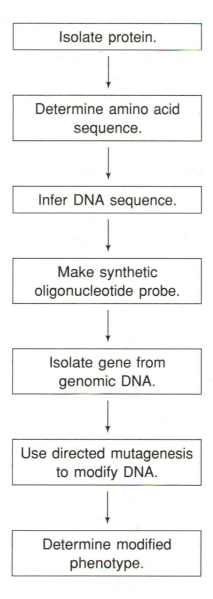

SOLVED PROBLEMS

The following problems have been chosen to represent the types of analyses regularly encountered by geneticists, since it is the belief of most genetics instructors that the act of doing genetics can be fairly accurately simulated by the working of such problems. As far as possible we have tried to relate the solutions to the principles contained in this book, but some cautions must be made. First, there are no "standard" or "machinelike" solutions to genetics problems; the principles can aid you but cannot do the job for you. Second (and this is probably just another way of stating the previous point) there is no substitute for common sense, and a little of it will go a long way in these solutions. Third, in the limited space available, it is not possible to illustrate all the principles listed.

1. Consider three yellow, round peas, A, B, and C. Each was grown into a plant and crossed to a plant grown from a green, wrinkled pea. One hundred peas issuing from each cross were sorted into phenotypic classes as follows:

> A: 51 yellow, round
> 49 green, round
> B: 100 yellow, round
> C: 24 yellow, round
> 26 yellow, wrinkled
> 25 green, round
> 25 green, wrinkled

What were the genotypes of A, B, and C? (Use gene symbols of your own choosing, making sure to define each one.)

SOLUTION

PRINCIPLE **11** concerns some of the standard phenotypic ratios in genetics, and here we see some examples.
The best clue is provided by the cross involving C, which
shows clearly the Mendelian segregation (**8**) of a
heterozygous gene pair determining yellow versus green,
and another independent gene pair (**9**) determining
round versus wrinkled.

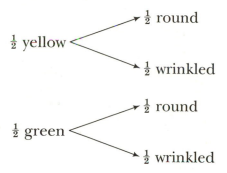

$\frac{1}{2}$ yellow $\nearrow$ $\frac{1}{2}$ round
$\searrow$ $\frac{1}{2}$ wrinkled

$\frac{1}{2}$ green $\nearrow$ $\frac{1}{2}$ round
$\searrow$ $\frac{1}{2}$ wrinkled

Cross C however, does not tell us which allele is dominant, but cross B does (see **6**): this cross of yellow,
round by green, wrinkled has only yellow, round
progeny; hence yellow (Y) is dominant to green (y) and
round (R) is dominant to wrinkled (r). Now we can write
cross C as:

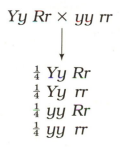

$$Yy\ Rr \times yy\ rr$$

$$\frac{1}{4}\ Yy\ Rr$$
$$\frac{1}{4}\ Yy\ rr$$
$$\frac{1}{4}\ yy\ Rr$$
$$\frac{1}{4}\ yy\ rr$$

Cross B must have been

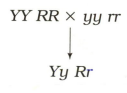

$$YY\ RR \times yy\ rr$$

$$Yy\ Rr$$

243

Cross A shows segregation (**8**) of only one gene pair as all progeny are round. Hence the cross must have been

$$Yy\ RR \times yy\ rr$$

$$\downarrow$$

$$\tfrac{1}{2}\ Yy\ Rr$$
$$\tfrac{1}{2}\ yy\ Rr$$

2. The accompanying pedigree concerns a certain rare disease X, that is incapacitating but not lethal.

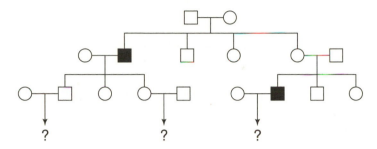

a. Determine the mode of inheritance of this disease.
b. Copy the pedigree into your answer book and alongside each individual write its genotype according to your proposed mode of inheritance.
c. If you were the family's doctor, how would you advise the three couples marked "?" about the likelihood of an affected child.

SOLUTION

Since this is a rare disease we can assume that people marrying into the family do not carry the gene for the disease (see the population principle **111**). Our main clue is provided by the fact that only men show the disease, but women, such as the last one in generation II, can obviously be carriers (heterozygotes). This is a standard inheritance pattern for an X-linked recessive phenotype (**13**). One of the grandparents must have carried the gene, and it must have been the grandmother, because if the grandfather carried the gene he would have expressed it since X-linked genes have no counterpart on the Y chromosome. Now we can write the pedigree as follows, where D = normal allele, d = disease allele.

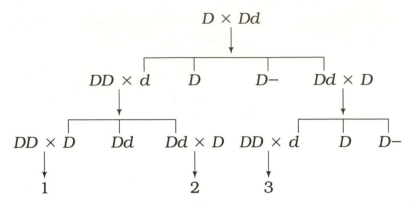

Notice that we can write *DD* or *D* for individuals marrying into the family. The *D–* means that we cannot be sure about the second allele.

Marriages

1. Obviously no risk of affected children

2. $\frac{1}{2}$ sons will have the disease
   $\frac{1}{2}$ daughters will be carriers
   Therefore, the chance of any child being affected is

   $\frac{1}{2}$ (for being a son) $\times$ $\frac{1}{2}$ (for being affected) $= \frac{1}{4}$

3. Sons will all be normal. Daughters will all be carriers. (No affected children)

3. The accompanying pedigree concerns a rare human disease.

  a. Copy the pedigree into your answer book, and alongside each individual write the genotype according to the most likely mode of inheritance.

  b. What would be the outcome of the cousin marriages: 1×9, 1×4, 2×3, and 2×8.

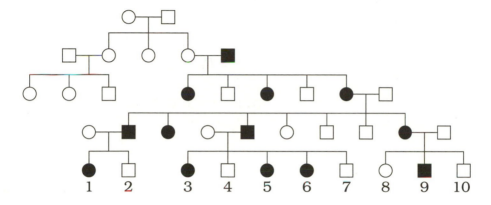

SOLUTION

In this pedigree we see that once introduced by the man in generation II, the disease appears in every generation. This is typical of a dominant allele **14** : the only way that a recessive could achieve the same result is if all people marrying into the family were carriers, and this is unlikely because the disease is rare. This dominant allele is most likely X-linked because men pass on the phenotype to daughters only, not to sons (see **13** for X linkage). If autosomal, both daughters and sons of affected men could be affected, and out of the total of 12 progeny of affected men in this pedigree, 3 would have been expected.

  a. Hence in the pedigrees, there are only four types, where P = disease, p = normal.

247

Affected men	P
Unaffected men	p
Affected women	Pp
Unaffected women	pp

NOTE: The only way a homozygous PP woman could arise is from a mating of two affected parents.

b. 1×9 $Pp \times P \longrightarrow$ ♀♀ $\frac{1}{2}$ PP, $\frac{1}{2}$ Pp
                       ♂♂ $\frac{1}{2}$ P , $\frac{1}{2}$ p

1×4 $Pp \times p \longrightarrow$ ♀♀ $\frac{1}{2}$ Pp, $\frac{1}{2}$ pp
                       ♂♂ $\frac{1}{2}$ P , $\frac{1}{2}$ p

2×3 $p \times Pp \longrightarrow$ same as 1×4.

2×8 $p \times pp \longrightarrow$ all normal.

4. The leaves of pineapples can be classified into three types: spiny, spiny-tip, and piping (nonspiny). In crosses made between pure strains the following results appeared:

PARENTAL PHENOTYPES	F_1 PHENOTYPES	F_2 PHENOTYPES
(a) spiny-tip × spiny	spiny-tip	3 spiny-tip: 1 spiny
(b) piping × spiny tip	piping	3 piping: 1 spiny-tip
(c) piping × spiny	piping	12 piping: 3 spiny-tip: 1 spiny

a. Using alphabetical letters, define gene symbols and explain these results in terms of the genotypes produced and their ratios.

b. Using your model from above give the phenotype ratios you would expect if you crossed: (a) the F_1 progeny from piping × spiny to the spiny parental stock; (b) The F_1 progeny of piping × spiny to the F_1 progeny of spiny × spiny-tip.

SOLUTION

The clue is provided by cross c in which the F_2 is obviously a modified Mendelian ratio (modified $9:3:3:1$). Modified Mendelian ratios reveal various kinds of gene interaction (**38**) so let us see which kind is operating in cross c and then work back to crosses a and b.

All $9:3:3:1$ ratios have the form:

9    $A-\ B-$

3    $A-\ bb$

3    $aa\ B-$

1    $aa\ bb$

so we can immediately infer that the spiny appearance is produced by the genotype $aa\ bb$. We can also infer that piping is produced by the $A-\ B-$ class and one other: since A and B are arbitrary symbols let us choose $A-\ bb$ as the other piping class. Automatically $aa\ B-$ is spiny tip. In summary:

$$\left.\begin{array}{l} A-\ B- \\ A-\ bb \end{array}\right\} = \text{piping}$$
$$aa\ B- = \text{spiny tip}$$
$$aa\ bb\ = \text{spiny}$$

The pure-breeding parents must have been $AA\ BB$ (piping) and $aa\ bb$ (spiny), and the F_1 was $Aa\ Bb$ (piping).

Turning to cross a, we see a single gene pair F_2 ratio of $3:1$ (see **11**) for spiny tip to spiny, and no piping individuals at all in any generation. Hence the pure-breeding parents must have been $aa\ BB$ (spiny tip) and $aa\ bb$ (spiny); the F_1 was $aa\ Bb$ (spiny tip) and the F_2 was $\frac{3}{4}$ $aa\ B-$ (spiny tip) and $\frac{1}{4}$ $aa\ bb$ (spiny). Using similar logic in line b we see that there is also a single gene F_2 ratio (**11**), and since there are no spiny individuals there is no b allele evident. Hence the parents were $AA\ BB$ (piping) and $aa\ BB$ (spiny tip), the F_1 was $Aa\ BB$ (piping), and the F_2 is $\frac{3}{4}$ $A-\ BB$ (piping) and $\frac{1}{4}$ $aa\ BB$ (spiny tip).

Now we can predict the progeny of the crosses in part b above:

1. The cross is *Aa Bb* × *aa bb*. Principle **24** tells us that the independence of the gene pairs allows us to predict that the gametes from the dihybrid will be $\frac{1}{4}\,AB:\frac{1}{4}\,Ab:\frac{1}{4}\,aB:\frac{1}{4}\,ab$, and since the *aa bb* individual produces only *ab* gametes, the $\frac{1}{4}:\frac{1}{4}:\frac{1}{4}:\frac{1}{4}$ ratio will be reflected in the phenotypic ratio.

2. The cross is *Aa Bb* × *aa Bb*. Taking the gene pairs one at a time we know that the progeny of a cross *Aa* × *aa* will be $\frac{1}{2}\,Aa:\frac{1}{2}\,aa$ (**8**). Furthermore, the progeny of a cross *Bb* × *Bb* will be $\frac{3}{4}\,B-:\frac{1}{4}\,bb$ (**11**). Since the gene pairs are independent, we can represent the progeny as follows:

$$\frac{1}{2}\,Aa \diagdown \begin{array}{l} \tfrac{3}{4}\,B- = \tfrac{3}{8}\,\text{piping} \\ \tfrac{1}{4}\,bb = \tfrac{1}{8}\,\text{piping} \end{array}$$

$$\frac{1}{2}\,aa \diagdown \begin{array}{l} \tfrac{3}{4}\,B- = \tfrac{3}{8}\,\text{spiny tip} \\ \tfrac{1}{4}\,bb = \tfrac{1}{8}\,\text{spiny.} \end{array}$$

Hence, in total there are

$$\tfrac{1}{2}\,\text{piping}:\tfrac{3}{8}\,\text{spiny tip}:\tfrac{1}{8}\,\text{spiny.}$$

Formally this system is a case of dominant epistasis of the *A* gene over the spiny tip *B* and spiny *b* alternatives, but as seen above the problem can be solved without this interpretation.

5. In Petunias, the normal purple colour is a result of a mixture of two pigments, red and blue, synthesized in separate biochemical pathways, as shown in the diagram, where white = colourless.

Pathway I

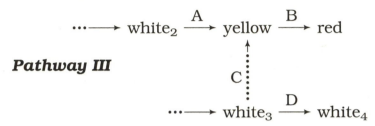

Red is formed through a yellow intermediate that normally does not reach detectable levels.

A third pathway involving only white compounds normally does not affect the blue and red pathways, but if one of its intermediates (white$_3$) should increase in concentration, it can be converted to the yellow intermediate of the red pathway.

In the diagram, A through E represent enzymes and their corresponding genes, all unlinked.

Assuming that wild-type alleles are dominant and represent enzyme function, and that recessive alleles are mutant and represent lack of enzyme function, which gene pairs, *in dihybrid crosses,* will give:

a.   9 purple : 3 green : 4 blue?
b.   9 purple : 3 red   : 3 blue : 1 white?
c.   13 purple : 3 blue?
d.   9 purple : 3 red   : 3 green : 1 yellow?

(NOTE: Blue mixed with yellow makes green; assume that no mutations are lethal.)

SOLUTION

This problem illustrates well the tight relationship between genetics and biochemistry. We shall see that it is the intricacies of the biochemical pathways and their enzymes that form the basis of many of the complex gene "interactions" revealed by modified Mendelian ratios (**38**).

Since the ratios mentioned are all modified two gene-pair ratios (**11**) we have to pick the enzymes two at a time to see which ones will give each ratio.

a.   In this ratio it is simple to see that the colour blue must be obtained by the presence of blue pigment and the absence of all others. This is obtained when enzyme A is deficient as when encoded by a mutant allele. What about green? There is no green pigment in the pathway shown, but green can be produced by mixing blue and yellow, a situation that would prevail if enzyme B is deficient. Hence the dihybrid in this case must have been *Aa Bb* where *a* and *b* represent the mutant alleles coding for enzymes A and B respectively, and *A* and *B* represent normal

253

functional alleles. We can see that the $9:3:4$ ratio is generated automatically as follows:

9   $A-$ $B-$ purple (= blue + red)

3   $A-$ bb green (= blue + yellow)

3   aa $B-$ blue

1   aa bb blue

Notice that the reason that aa bb is blue and not green is that enzyme A is earlier in the synthesis of red pigment than B, so the genotype aa makes the B genotype irrelevant.   **37 45**

b.   A similar logic can be used here to see how the colour red can be produced. Obviously to get red by itself blue must be eliminated, so we can immediately identify E as one of the genes we need and the dihybrid Aa Ee. As before blue is obtained by a deficiency of A. In summary:

9   $A-$ $E-$ purple

3   $A-$ ee red

3   aa $E-$ blue

1   aa ee white

In this case the double homozygous recessive is white because no pigment is produced at all— whiteness is really a result of the sun's white light reflecting off the cells of the petal, with none being absorbed by pigment.

c. In this case, *A* again is the gene whose mutated allele will produce blue. Trial and error will reveal that *D* is the other gene we are looking for because when deficient, the compound white$_3$ builds up and is converted to yellow and hence bypassing a blockage caused by *aa*.

9   *A– D–*   purple

3   *A– dd*   purple

3   *aa D–*   blue

1   *aa dd*   purple

Note that here *dd* is acting as a suppressor of *aa*.

d.   *Bb Ee* (similar logic).

This question illustrates well the combination of principles, common sense, and trial and error that must be brought to bear on most pieces of genetic analysis.

6. In *Drosophila* the allele *b* gives black body (normally brownish); at a separate gene the allele *wx* gives waxy wings (normally nonwaxy); and at a third gene the allele *cn* gives cinnabar eyes (normally red). An individual heterozygous for these three genes was test-crossed and 1,000 progeny were classified as follows: 5 normal; 6 black, waxy, cinnabar; 69 waxy, cinnabar; 67 black; 382 cinnabar; 379 black, waxy; 48 waxy; and 44 black, cinnabar.

    a. Explain these numbers.
    b. Draw the alleles in their proper positions on the chromosomes of the triple heterozygote.
    c. If it is appropriate under your explanation, calculate interference.
    d. If the triple heterozygote were selfed, what proportion of progeny would be black, waxy, cinnabar?

SOLUTION

We are told that the cross made was of the following type:

$$b^+b \; wz^+wx \; cn^+cn \times bb \; wxwx \; cncn$$

(This type of cross is discussed in **27**).

Initially we do not know if the genes are linked, and if so in what order, but the genes are simply listed in the order described.

As in many problems, it is helpful to rewrite the data in your own way to reinforce their significance. Since it is clear that the problem will involve recombination in the triple heterozygote we can benefit from listing the progeny as gamete genotypes from that individual—the gametic contribution of the fully recessive tester can be ignored because it does not provide any information.

+	+	+	5
b	wx	cn	6
+	wx	cn	69
b	+	+	67
+	+	cn	382
b	wx	+	379
+	wx	+	48
b	+	cn	44

It is now clear that the gene pairs are not segregating independently (**9**) because then we would expect $2 \times 2 \times 2 = 8$ equal classes. The numbers show that in the heterozygote we were dealing with a pair of homologues one of which carried the *cn* allele, and the other the *b* and *wx* alleles. But we still cannot draw these homologues because we do not know gene order.

There are two ways of deriving gene order. We could calculate all the three possible recombinant frequencies and draw the chromosome map right away (**26 27**). Another way of obtaining order is to look at the constitution of the double-crossover chromatids. These must be the first two types because they are the least frequent.

We then compare these with the parental chromosomes (the most frequent classes) with regard to allele conformation. Here we see that both the double crossovers and the parentals have the $b^+ wx^+$ or $b\ wx$ arrangements, but differ with regard to the cn conformation. Hence the cn locus must be in the middle, flanked by the b and wx loci. The double crossovers arise as follows:

$$
\begin{array}{ccc}
b & + & wx \\
\hline
X & X & \\
\hline
+ & cn & + \\
\end{array}
\quad \longrightarrow \quad
\begin{array}{ccc}
b & cn & wx \\
\hline
& \text{and} & \\
\hline
+ & + & + \\
\end{array}
$$

Recombinant frequencies (RF) will give us map distances. For the b and cn loci, the RF $= [(5 + 6 + 48 + 44) \div 1000] \times 100 = 10.3$ cM. For the cn and wx loci, the RF $= ([5 + 6 + 69 + 67) \div 1000] \times 100 = 14.7$ cM. Hence the chromosome map is

$$
\begin{array}{ccc}
b & cn & wx \\
\vdash\!\!\!\!\!\!\!-\!\!\!-\!\!\!-\!\!\!-\!\!\!+\!\!\!-\!\!\!-\!\!\!-\!\!\!-\!\!\!-\!\!\!+ \\
\text{10.3 cM} & \text{14.7 cM} \\
\end{array}
$$

To see the level of interference, we apply the formula in (**27**).

$$
I = 1 - \frac{11}{(.103 \times .147 \times 1000)} = 0.27 \text{ or } 27 \text{ percent.}
$$

Section d asks for the proportion of $bb\ wx\ wx\ cn\ cn$ progeny in a self of the triple heterozygote. This genotype can be produced only by the union of a $b\ wx\ cn$ egg with a $b\ wx\ cn$ sperm. We know that the frequency of such gametes will be 0.006 from an inspection of the gametes in the testcross. Hence the answer will be $0.006 \times 0.006 = 0.000036$ or 0.0036 percent.

7. A yeast plasmid carrying the yeast $leu2^+$ gene is used to transform nonrevertible haploid $leu2^-$ yeast cells. Several leu^+ transformed colonies appear on a leucineless medium. Thus presumably $leu2^+$ DNA entered the recipient cells, but now you have to decide what happened to it once inside. Crosses of transformants to $leu2$ testers revealed that there are three types of transformants: A, B, and C, reflecting three different fates of the $leu2^+$ in the transformants. The results were

$Type\ A \times leu2^-$
$\longrightarrow \frac{1}{2}\ leu^-$
$\quad\quad \frac{1}{2}\ leu^+,\ \times$ standard $leu2^+ \longrightarrow \frac{3}{4}\ leu^+$
$\quad\quad\quad\quad\quad\quad\quad\quad\quad\quad\quad\quad\quad\quad\quad \frac{1}{3}\ leu^-$

$Type\ B \times leu2^-$
$\longrightarrow \frac{1}{2}\ leu^-$
$\quad\quad \frac{1}{2}\ leu^+,\ \times$ standard $leu2^+ \longrightarrow$ 100% leu^+
$\quad\quad\quad\quad\quad\quad\quad\quad\quad\quad\quad\quad\quad\quad\quad$ 0% leu^-

$Type\ C \times leu2^- \longrightarrow$ 100% leu^+.

What three different fates of the $leu2^+$ DNA are suggested by these results? Be sure to explain *all* the results under your hypotheses. Use diagrams if possible. (Question devised by Jeanette Leach)

SOLUTION

This question concerns the genetics of a haploid organism, yeast (see **10** for the relevant life cycle). We have to juggle several different principles to understand these data. Let's start with the easier ones first (This is always a good idea—experiments rarely can be understood fully in the sequence they are described).

Transformation (**87**) is a standard technique of creating genetically engineered organisms, and the kinds of results obtained here are commonly encountered.

Type B leu^+ transformants give a Mendelian single-gene ratio when crossed to a $leu2^-$ strain (see **8** —Mendel's first law). This suggests strongly that the $leu2^+$ gene has left the plasmid and inserted itself at the $leu2$ locus where it is behaving as a proper $leu2^+$ allele. This is confirmed by the cross to $leu2^+$, which produces all leu^+ progeny.

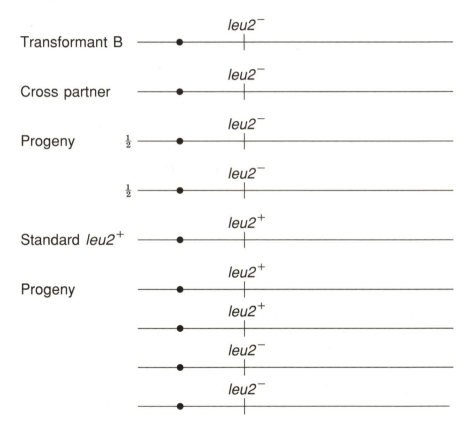

Type A transformants also show a 1 : 1 ratio in the initial cross to *leu2⁻*, representing a single gene segregation. But the *leu⁺* progeny of that cross do not behave as alleles of the *leu2⁺* locus when crossed to *leu2⁺*. In fact the 3 : 1 ratio in haploids is indicative of two independent segregating loci **9** . This is explained if the *leu2⁺* insert is on a separate chromosome independent of the normal *leu2⁺* locus. The diagram spanning these two pages shows the usual *leu2* chromosome (left) and the chromosome with the new insert (right).

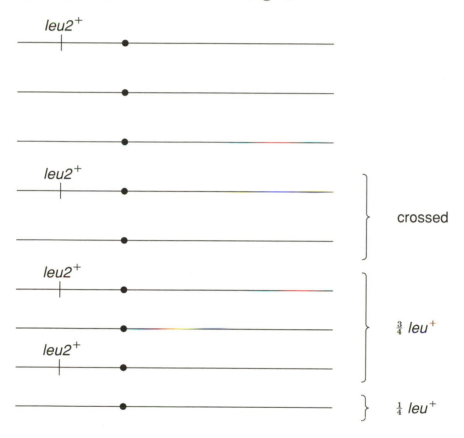

Type C transformants transmit the *leu*⁺ phenotype to all the progeny. This is a kind of uniparental inheritance and suggests an extrachromosomal basis of the *leu*⁺ phenotype (see **30**), precisely what would be expected if the *leu*⁺ phenotype was caused by persistence of the transforming plasmid in high copy number so that each meiotic product received a copy. Of course, here the *leu2*⁺ gene does not integrate at all.

8. In *Drosophila*, the eye phenotype "star" is caused by recessive mutations (s) mapping to one location on the second chromosome. This region is flanked to the left by the locus A/a and to the right by B/b, as follows.

A/a *star* B/b

A set of six independently induced star mutations is each made homozygous with both $AA\ BB$ and $aa\ bb$ constitutions and intercrossed to study complementation at the star locus. The results were as follows, where $+$ = wild eye, and s = star eye, both being phenotypes of the F_1.

	AA ss BB					
aa ss bb	1	2	3	4	5	6
1	s	+	s	s	+	(+)
2		s	+	(+)	s	+
3			s	s	+	+
4				s	+	+
5					s	+
6						s

a. How many cistrons are at the star location, and which mutant sites are in each?

b. The heterozygotes in brackets were allowed to produce gametes and in both cases star$^+$ recombinant gametes were identified. These were then tested for the flanking marker conformation which was $a\ B$ in both the 1/6 heterozygote and the 2/4 heterozygote. Order the cistrons in relation to the A/a and B/b loci.

Although all the *star* mutations produce the same effect and map to approximately the same locus, they are obviously not all of the same cistron (gene). This is shown by the fact that in some combinations seen in the table they show complementation (**53**) and if they were all at the same cistron there would be no complementation at all.

The best way of determining which mutant site is in which cistron **54** is to group those that do not complement. Mutant 1 does not complement with 3 or 4 so 1, 3, and 4 are all in one cistron. Similarly 2 and 5 are in another cistron, and 6 alone is in still another. Hence there are three cistrons — presumably three virtually adjacent genes of related function. A mutation in either, gives the same phenotype, presumably because of the related function — possibly three consecutive steps (**37**) in the synthesis of the same eye product.

The synthesized *A s B/a s b* heterozygotes constitute the form of the complementation test in a diploid organism such as *Drosophila*. But these same heterozygotes can be used to allow a recombination analysis (**26**). By selecting *star*$^+$ recombinants we detect crossovers occurring between the mutant sites. In the $\frac{1}{6}$ heterozygote these *star*$^+$ recombinants were *aB* in constitution, hence mutant site 6 must have been to the left of 1 as follows:

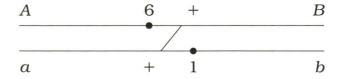

The same reasoning tells us that 4 is to the left of 2. As we have used representatives of all three cistrons we now know the order must be:

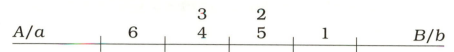

9. Five Hfr strains *A* to *E* are derived from a single F⁺ strain of *E. coli*. The following chart shows the time of entry of the first five markers into an F⁻ strain when each is used in an interrupted conjugation experiment.

Hfr Strain

A	B	C	D	E
mal^+ (1)	ade^+ (13)	pro^+ (3)	pro^+ (10)	his^+ (7)
str^s (11)	his^+ (28)	met^+ (29)	gal^+ (16)	gal^+ (17)
ser^+ (16)	gal^+ (38)	xyl^+ (32)	his^+ (26)	pro^+ (23)
ade^+ (36)	pro^+ (44)	mal^+ (37)	ade^+ (41)	met^+ (49)
his^+ (51)	met^+ (70)	str^s (47)	ser^+ (61)	xyl^+ (52)

a. Draw a map of the F⁺ strain showing the positions of all genes and their distances apart in minutes.
b. Show the insertion point and orientation of the F plasmid in each Hfr.
c. In using each of these Hfr strains, state which gene you would select to obtain the highest proportion of Hfr exconjugants.

SOLUTION ───────────────────────────────

In bacteria there are several ways of obtaining a chromosome map (**28**) and this question involves the interrupted conjugation method. During the process of chromosome transfer the cells are sampled for the appearance of donor (Hfr) genes. Five different Hfrs are used in this experiment. Once the genes have entered they are part of a merozygote and must combine with the F⁻ chromosome (**23**) to become part of the F⁻ genome upon cell division.

Starting with Hfr *A*, we see that the *mal*⁺ gene enters at 1 minute from the beginning of the experiment, followed 10 minutes later by *str*ˢ, and 5 minutes after that by *ser*⁺, and so on. Hence, a map emerges of that part of the chromosome using minutes as map units

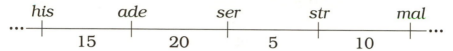

Looking at Hfr *B* we can expand this view as follows:

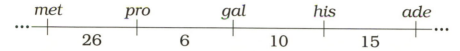

Extending this process, we can only make sense of all the data if the map is a circle as shown in the following diagram:

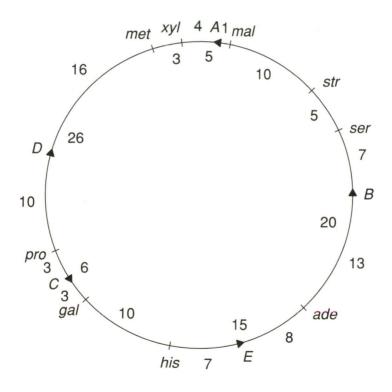

The position and orientation of the inserted *F* plasmid (F factor) is revealed by the first gene to enter. In the case of Hfr *A mal* is first, so the *F* insertion is between *mal* and *xyl*, oriented counter-clockwise. It is known that part of the *F* plasmid enters at the head of the transferred molecule, but the other part is not transferred until the very tail end. Thus to select for complete transfer of the Hfr type we need to select the last marker to enter. They are:

A	*xyl*
B	*ser*
C	*gal*
D	*met*
E	*ade*

10. In a generalized transducing system using P1 phage the donor was: pur^+ nad^+ pdx^- and the recipient was: pur^- nad^- pdx^+. The donor allele pur^+ was initially selected after transduction and 50 pur^+ transductants were then scored for the other alleles present. The results were as follows:

Genotypes		Number of colonies
nad^+	pdx^+	3
nad^+	pdx^-	10
nad^-	pdx^+	24
nad^-	pdx^-	13
TOTAL		50 colonies

a. What is the cotransduction frequency for pur and nad?
b. What is the cotransduction frequency for pur and pdx?
c. Which of the unselected loci is closest to pur?
d. Are nad and pdx on the same or on opposite sides of pur? Explain. (Draw the exchanges needed to produce the various transformant classes under either order to see which requires the minimum number to produce the results obtained.)

SOLUTION ——————————————————————————————

The approach taken to finding the relative map positions of these bacterial genes was to transform (**87**) using one gene and find how often other genes are transformed at the same time—presumably reflecting location of both genes on the same fragment and insertion of the fragment by recombination into the chromosome of the recipient (**23**).

269

Cotransduction for donor alleles pur^+ and nad^+ is seen in $10 + 3 = 13$ out of 50 colonies, or 26 percent, whereas cotransformation for donor alleles pur^+ and pdx^- is seen in $10 + 13 = 23$ out of 50 or 46 percent. Hence, the pdx gene must be closest to pur because they are processed simultaneously the greater proportion of the time.

The pdx and nad loci must be on opposite sides of pur:

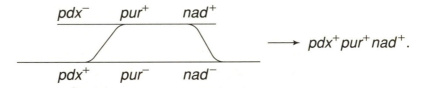

In this conformation all the four transformant genotypes can be produced by the usual merozygote double crossovers, for example,

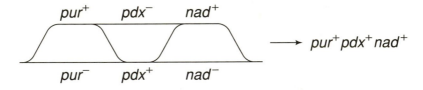

Whereas, if both on the same side with pdx closest, a quadruple crossover is needed to produce one class—pdx^+ pur^+ nad^+

11. In *Neurospora* the markers *ad-3* and *pan-2* are auxo-trophic mutations located on chromosomes I and VI respectively. An unusual *ad-3* line arose in the labo-ratory which gave the following results:

	Ascospore appearance	RF between *ad-3* & *pan-2*
1. Normal *ad-3* × normal *pan-2*	all black	50%
2. Abnormal *ad-3* × normal *pan-2*	about $\frac{1}{2}$ black & $\frac{1}{2}$ white (inviable)	1%

3. Of the black spores from cross 2, about half were completely normal and half repeated the same behaviour as the original abnormal *ad-3* strain.

Explain all three results with the aid of clearly labelled diagrams.

NOTE: In *Neurospora,* ascospores with extra chro-mosomal material survive and are the normal black colour, whereas ascospores lacking any chromo-somal region are white and inviable.

Line 1 confirms the normal expected behavior of these genes—independent assortment with a recombinant fraction of 50 percent.

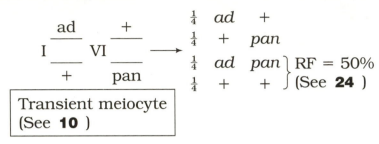

Line 2 gives us two important clues. First the RF has dropped almost to zero—in other words, the normally independent loci now appear to show linkage. The simplest and best explanation for this is that there has been a translocation (**64**) between chromosomes I and VI in the abnormal strain. Second there is 50 percent abortion of the meiotic products (effectively the equivalent of semisterility in a plant), precisely what is expected in a translocation. The translocation would look something like this:

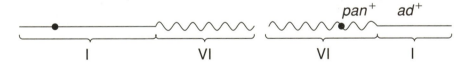

The meiotic configuration would be:

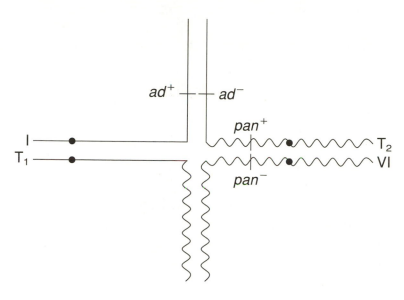

Segregations:

> *alternate:* $T_1 + T_2$ $ad^-\ pan^+$
> I + VI $ad^+\ pan^-$

> *adjacent:* $\left.\begin{array}{l}T_1 + VI \\ I + T_2\end{array}\right\}$ abort because
> duplication-deletion types

Of the viable spores, we see, as borne out in cross 3, that half are normal and half are translocation types of the original abnormal *ad-3* constitution and expected to repeat its behavior.

12. In a certain plant there is a controversy about the type of chromosome pairing seen in allotetraploids between two geographical races. It is known that chromosome association is by pairs, but three hypotheses are put forward as about this occurs.

a. Pair formation is random.

b. Pairs only occur between chromosomes of the same race.

c. Pairs only occur between chromosomes of different races.

For a gene A/a, closely linked to its centromere, a cross is made:

Race 1: $AAAA$ × Race 2: $aaaa$

$$\downarrow$$

Allotetraploid: $AAaa$

The allotetraploid is then selfed. What ratio of phenotypes is expected under each hypothesis of chromosome pairing? Explain your answer.

SOLUTION ──────────────────────────

From principle **66** we see that there are several types of pairing in tetraploids, of which the most stable is shown in this system.

a. If pairing is random, the result obtained is shown in (**67**), in other words a $35 : 1$ ratio of A— to $aaaa$.

b. This pairing will be of the following type:

$$\frac{A}{A} \qquad \frac{a}{a}$$

and will produce gametes, all of which are diploids of genotype Aa. Thus *all* the progeny will be $AAaa$ and of A phenotype.

c. This pairing is:

$$\frac{A}{a} \qquad \frac{A}{a}$$

We can represent the gamete production one chromosome pair at a time as follows:

$$\frac{1}{2} A \left\langle \begin{array}{l} \frac{1}{2} A \longrightarrow \frac{1}{4} AA \\ \frac{1}{2} a \longrightarrow \frac{1}{4} Aa \end{array} \right.$$

$$\frac{1}{2} a \left\langle \begin{array}{l} \frac{1}{2} A \longrightarrow \frac{1}{4} Aa \\ \frac{1}{2} a \longrightarrow \frac{1}{4} aa \end{array} \right.$$

Although many progeny genotypes will be produced, the phenotypic ratio is easy to calculate:

$$aaaa = \tfrac{1}{4} \times \tfrac{1}{4} = \tfrac{1}{16}$$
$$A\text{—} = 1 - \tfrac{1}{16} = \tfrac{15}{16}$$

In other words there will be a $15 : 1$ ratio. Thus in summary, the three possibilities produce very different ratios of $35 : 1$, $1 : 0$, and $15 : 1$.

13. Using tomatoes, an attempt was made to assign 5 recessive genes to specific chromosomes using trisomics. Each homozygous mutant (*2n*) was crossed to three trisomics, involving chromosomes 1, 7, and 10. From these crosses trisomic progeny (less vigorous) were selected. These trisomic progeny were backcrossed to the appropriate homozygous recessive, and *diploid* progeny were examined from these crosses. The results were as follows, where the ratio are of wild type: mutant.

Trisomic chromosome	Gene				
	d	*y*	*c*	*h*	*cot*
1	48 : 55	72 : 79	56 : 50	53 : 54	32 : 28
7	52 : 56	52 : 48	52 : 51	58 : 56	81 : 40
10	45 : 42	36 : 33	28 : 32	96 : 50	20 : 17

Which of the genes can be assigned, and to which chromosomes? (Explain your answer fully).

SOLUTION

The cross programme can be written out as follows using mutant *h* as an example:

$$hh \times \text{trisomic } 1 \longrightarrow \text{trisomic; } \times hh$$
$$hh \times \text{trisomic } 7 \longrightarrow \text{trisomic; } \times hh$$
$$hh \times \text{trisomic } 10 \longrightarrow \text{trisomic; } \times hh$$

The F_1 trisomics are crucial because if *h* is on one of the chromosomes 1, 7, or 10, these trisomics will be of the constitution $h^+ h^+ h$ and, as shown in **67** , a non-Mendelian ratio will be obtained. If the gene *h* is not on 1, 7, or 10, the chromosome that it is on will be of the conformation $h^+ h$ and a regular Mendelian segregation will be obtained in the cross to *hh*.

276

In the case of gene *h*, we see that Mendelian ratios are produced for chromosomes 1 and 7, but for chromosome 10 a 2 : 1 ratio is shown, suggesting that *h* is on chromosome 10.

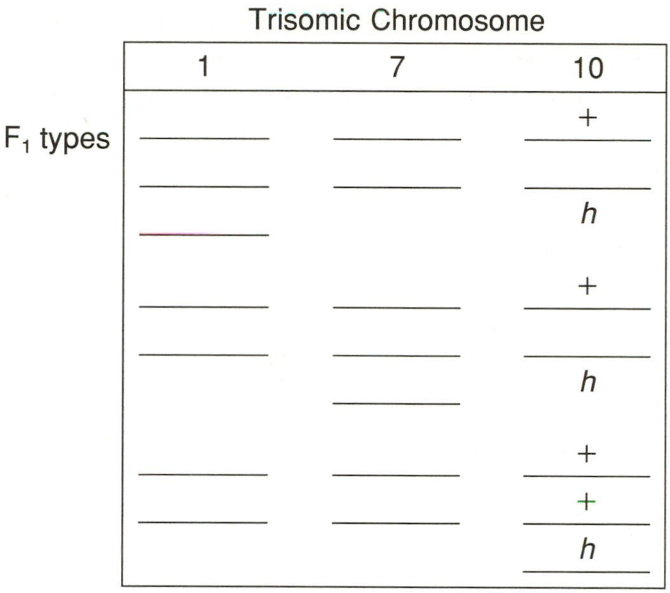

If the ++*h* trisomic homologues show one chromosome pair and one unpaired chromosome, we have the following gametes produced in the proportions shown:

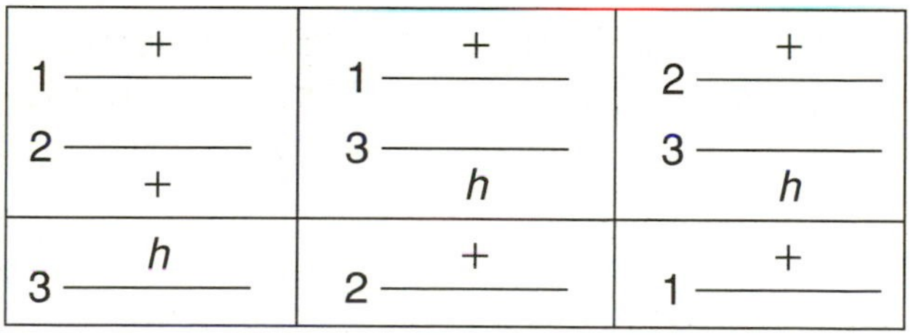

The second row shows the types that will produce diploid progeny (these are the types shown in the question table), and it is seen that a 2 : 1 ratio will hold for a cross to *hh*.

14. DNA studies are performed on a large family which shows a certain autosomal dominant disease of late onset (age approximately 40 years). A DNA sample from each family member is digested with the restriction enzyme Taq 1 and run on an electrophoretic gel. A southern analysis is then performed using a radioactive probe consisting of a portion of human DNA cloned in a bacterial plasmid. The autoradiogram is shown below, aligned with the family pedigree.

Pedigree:

Autoradiogram:

Migration

a. Analyze fully the relationship between the DNA variation, the probe DNA, and the gene for the disease. Draw the relevant chromosomal regions.

b. How do you explain the last son?

c. Of what use would these results be in counselling subsequent marriages of people from this family?

Again in this question we are asked to relate classical genetics to the situation at the molecular level. The question illustrates an important new technique useful in all organisms, but particularly in humans, called restriction-fragment-length polymorphism (RFLP) mapping (see **108**). The probe (**92**) has picked up an RFLP—two DNA types in the area homologous to the probe. The situation can be diagrammed as follows:

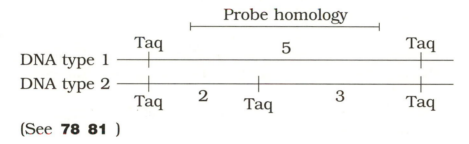

(See **78 81**)

More interesting, type 2 is associated, in this family, mainly with two individuals bearing the disease. But the last male shows us that the presence of the central Taq 1 site is not itself the cause of the disease. Rather, what we seem to have is a linkage relation of the disease gene with the type 2 region of DNA, as follows:

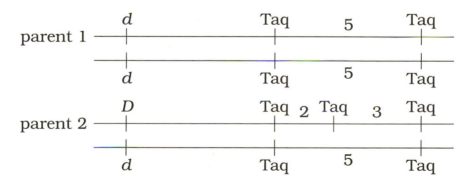

A crossover between the gene and the RFLP would produce two reciprocal gametic types of which the last male is derived from one

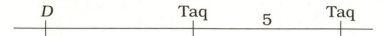

RFLPs probably represent mostly neutral DNA variation (**108**). Nevertheless they are very common—virtually any cloned piece of DNA will pick up an RFLP in a small population. In humans in particular, they constitute highly useful chromosomal landmarks to which other genes can be mapped.

In the present family the RFLP is useful in diagnosing the disease before onset (i.e., before having children).

NOTE: In this analysis the RFLP is merely being used as a marker (**47**).

15. A linear phage chromosome is labeled at both ends with ^{32}P and digested with restriction enzymes. EcoRl produces fragments of sizes 2.9, 4.5, 6.2, 7.4 and 8.0 kb. An autoradiogram developed from a Southern blot of this digest shows radioactivity associated with the 6.2 and 8.0 fragments. BamHl cleaves the same molecule into fragments of size 6.0, 10.1 and 12.9, and the label is found associated with the 6.0 and 10.1 fragments. When EcoRl and BamHl are used together, fragments of sizes 1.0, 2.0, 2.9, 3.5, 6.0, 6.2 and 7.4 kb are produced.

    a. Draw a restriction enzyme target site map of this molecule, showing relative positions and distances apart.

    b. A radioactive probe made from a cloned phage gene *X* is added to Southern blots of single enzyme digests of phage DNA. The autoradiograms showed hybridization associated with the 4.5, 10.1, and 12.9 kb fragments. Draw in the approximate location of gene *X* on the restriction map.

SOLUTION

Eco Digest (See (**78**):

The Southern (**91**) shows that the 6.2 and 8.0 kb fragments are at the ends of the phage chromosome. The order of the others is not deducible.

BamHl Digest:

The 6.0 and 10.1 kb fragments are the ends.

Double Digest:

- Eco 7.4 and 2.9 contain no Bam sites.
- Eco 8 cut by Bam to 6 and 2.
- Eco 4.5 cut by Bam to 3.5 and 1.

Taken together, these data show:

Eco Sites:

Bam Sites:

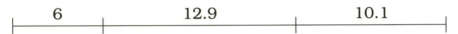

Restriction Map: (82)

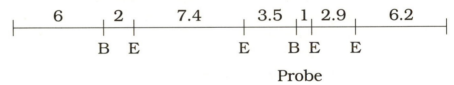

Probe

Using gene X as a probe (**92**) picks up fragments 4.5, 10.1, and 12.9 in the single digests, therefore the gene's position is within the Eco 4.5 fragment.

16. A fragment of human genomic DNA for gene *P* is excised from a λ vector using Hind III, and the ends of the gene fragment (GP) labelled with ^{32}P. The fragment is initially digested with Bam, giving two fragments, of 1.0 kbp and 10 kbp. The 10 kbp fragment is *partially* digested with Hpa II and the products electrophoresed, giving the results below.

Autoradiogram of Hpa II digest of GP fragment:

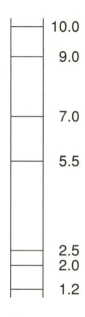

10.0

9.0

7.0

5.5

2.5
2.0

1.2

a. Map the genomic fragment, indicating the Bam site and Hpa II sites.

b. The cDNA for gene P can be cleaved by Hpa II into only two fragments. These are both labeled and each used as a ^{32}P-probe to investigate fragments from a complete digest of GP using Hpa II and Bam enzymes, on Southern Blots (below).

	Southern Blots of complete digest	
Electrophoresis of products of Bam & Hpa II complete digest of GP.	Probe 1	Probe 2
	^{32}P cDNA fragment 1	^{32}P cDNA fragment 2

Electrophoresis of products of Bam & Hpa II complete digest of GP (Ethidium stained):
3.0, 2.0, 1.5, 1.2, 1.0, 0.8, 0.5

Probe 1 autoradiogram (^{32}P cDNA fragment 1):
3.0, 1.2

Probe 2 autoradiogram (^{32}P cDNA fragment 2):
1.5, 1.0

Ethidium stained — Autoradiograms

Draw a diagram comparing genomic DNA with cDNA. Indicate the restriction sites. Explain the differences.

Mark the Hpa II site on the cDNA, and the orientation of the two Hpa fragments.

SOLUTION ───────────────────────────────────

a. *Initial Digest* **78 81**

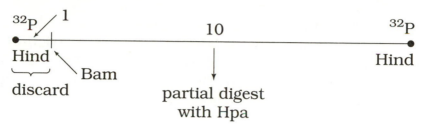

In the partial digest the key concept is that not all the Hpa sites are cut in all DNA molecules. Hence we can visualize a mixture of lengths of radioactive pieces as follows:

Where H = Hpa

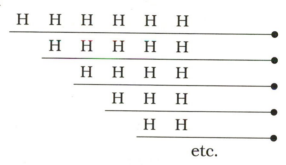

Therefore the differences in sizes allow us to map the Hpa sites directly

(10 − 9 =)
│   (9 − 7 =) etc. ⟶
↓   ↓

   1   2   1.5   3   .5 .8 1.2

H   H   H   H   H H H

b. The cDNA (see **90**) obviously has only one Hpa site—an interesting point that has to be explained later.

Hpa fragment 1 when used as a probe shows homology (**18 84 91**) to genomic restriction fragments 3.0 and 1.2, whereas fragment 2 shows homology to fragments 1.5 and 1.0. This homology can be diagrammed as follows:

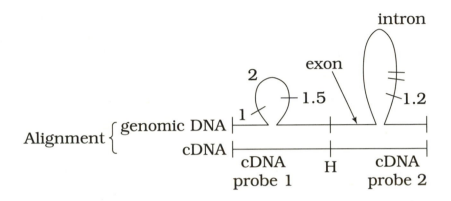

All the marked sites are Hpa sites.

The two regions of the gene not represented in the cDNA are evidently two introns that are excised during the synthesis of mRNA (**76**).

17. The gene for β-tubulin has been cloned from *Neurospora* and is available, carried in a λ vector. Now you wish to clone the same gene from the related fungus *Podospora*, but the cloning vector you wish to use is the *E. coli* plasmid shown below. List a step-by-step procedure for doing this.

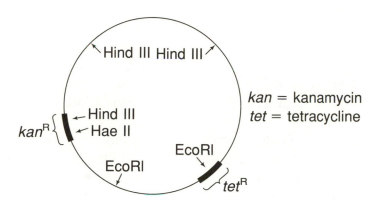

The availability of the equivalent *Neurospora* gene is very convenient because evolutionary conservation (**97**) makes it very likely that these two closely related fungi share considerable base-pair homology (**84**), especially for an important protein like tubulin, which constitutes cellular microtubules.

First a library **88** of the *Podospora* genome must be prepared using the vector shown. The restriction enzyme to use is Hae II because this has a unique cutting site which is in a selectable gene, *kan*^{R} (**86**).

Therefore, *Podospora* genomic DNA and the vector DNA are cut with Hae II and mixed and ligated (**85 86**). This population of molecules is then used to transform *E. coli* (**87**) and the cells are plated on tetracycline to select for successful transformants that bear the plasmid.

Transformants with an insert in *kan*^{R} should be *tet*^{R}*kan*^{S} (**86**) so tests are made to identify the *kan*^{S} clones. The *tet*^{R}*kan*^{S} clones are then probed with radioactive λ vector containing the *Neurospora* tubulin gene (**89**). Those colonies which show up on the autoradiogram are the ones that contain the *Podospora* gene of interest.

18. In a population of cattle, three different characters showing continuous distribution were measured and the following variances were calculated:

	Characters		
	Shank length	Neck length	Fat Content
Phenotypic variance	310.2	730.4	106.0
Environmental variance	248.1	292.2	53.0
Additive genetic variance	46.5	73.0	42.4
Dominance genetic variance	15.6	365.2	10.6

a. Calculate broad- *and* narrow-sense heritabilities for these characters.

b. In the population of animals studied, which character would respond best to selection?

c. A project was undertaken to decrease mean fat content in the herd. The mean fat content was 10.5 percent. Animals of 6.5 percent fat content were interbred as parents of the next generation. What mean fat content is expected in the descendants of these animals.

a. The formula for broad-sense heritability is in **104** . The calculations are:

$$\text{shank } (46.5 + \ \ 15.6)/310.2 = 0.2$$
$$\text{neck } \ \ (73.0 + 365.2)/730.4 = 0.6$$
$$\text{fat } \ \ \ \ \ (42.4 + \ \ 10.6)/106.0 = 0.5$$

Narrow-sense heritability is defined in **105** . The calculations are:

$$\text{shank } 46.5/310.2 = 0.15$$
$$\text{neck } \ \ 73.0/730.4 = 0.10$$
$$\text{fat } \ \ \ \ \ 42.4/106.0 = 0.40$$

b. The character with the highest narrow-sense heritability would respond best because this is precisely what is measured by this value. Hence, fat content is the answer.

c. This project is based on the theory in **105** . The formula can be substituted as follows:

$$0.4 = \frac{y}{10.5 - 6.5}$$

Therefore,

$$y = 0.4 \times 4.0 = 1.6$$

Therefore, the new mean will be

$$10.5 - 1.6 = 8.9.$$

19. In an extremely large experimental population of fruitflies, the T/t locus is thought to be in Hardy-Weinberg equilibrium. It is known that the proportion of recessive homozygotes is 4 percent.

a. What are the predicted p and q values?

b. What are the predicted proportions of dominant homozygotes and heterozygotes?

c. A decision is made to test the idea that the population is in Hardy-Weinberg equilibrium, by making random matings between individuals with the dominant phenotype. Under the Hardy-Weinberg hypothesis, what proportion of such matings are expected to give rise to some progeny of genotype tt?

d. In another experiment, artificial selection was imposed against individuals of genotype tt so that their fitness alone is reduced to 0.5. What will be the values of p and q after one generation of such selection?

e. If fitness of tt were reduced to zero, what will be p and q after 50 generations of selection?

SOLUTION ————————————————————————

a. Principle **110** shows us the necessary formula to calculate p and q. The first thing to realize is that 4 percent is the value of q^2.

Hence,

$$q^2 = 0.04, \therefore q = 0.2$$
$$p = 1 - q, \therefore p = 0.8$$

b. TT would be $p^2 = (0.8)^2 = 0.64$, Tt would be $2pq = 2 \times 0.8 \times 0.2 = 0.32$.

c. The first two parts were answered on the *assumption* that the population was in Hardy-Weinberg equilibrium, now we proceed to make predictions that can be tested. Principle **112** shows the mating structure of a randomly mating population in Hardy-Weinberg equilibrium. We need parts of that grid to answer the question. First note the ratio of TT to Tt genotypes is $(\frac{2}{3}):(\frac{1}{3})$ $(= 0.64:0.32)$. Now we can draw the following grid of the relevant matings:

	$TT\ \frac{2}{3}$	$Tt\ \frac{1}{3}$
$TT\ \frac{2}{3}$	No *tt* offspring	No *tt* offspring
$Tt\ \frac{1}{3}$	No *tt* offspring	$(\frac{1}{3})^2 = \frac{1}{9}$

Hence 1/9 of all matings between randomly selected dominants are expected to have *tt* progeny. The experiment can be done and tested against this expectation based on Hardy-Weinberg equilibrium.

d. The best way to set up the selection calculation for one generation of selection is as follows (**113**):

Genotype	Frequency	Allelic contribution to next generation	
		T	t
TT	0.64	0.64	0
Tt	0.32	0.16	0.16*
tt	0.04	0	0.02**

$$0.80 \qquad 0.18$$
$$0.98$$

*Because *Tt* contribute equal numbers of *T* and *t*
**Because halving their fitness means halving their contribution

Hence, new

$$p = 0.80 \div 0.98 = 0.82$$
$$q = 0.18 \div 0.98 = 0.18$$

e. This calculation requires the chart on (**113**). We see that when fitness = 0 (full selection against *tt*), if the allele frequency of *t* is expressed as a fraction with 1 as the numerator, then the denominator is increased by 1 in each generation of selection. Since q starts as $0.2 = \frac{1}{5}$, then 50 generations of selection would take us to $\frac{1}{5+50} = \frac{1}{55}$. Hence q would be *0.018* (Note how inefficient this selection is).

20. A population of snails was studied for possible electrophoretic alleles of the alcohol dehydrogenase gene. Snail extracts from many individuals were run on electrophoretic gels, and then the gels were stained for alcohol dehydrogenase. It turned out that all the snails gave one of three patterns of bands on the gels, and these are shown below, together with the number of snails in each class.

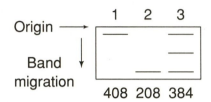

Classes 1 and 2 bred true, whereas class 3 gave rise to all three types when selfed.

a. Invent allele symbols and give the genotypes of the three classes.

b. What are the allele frequencies in the population?

c. Is the population in Hardy-Weinberg equilibrium?

d. If the inbreeding coefficient is 0.2, what heterozygote frequency would result with these p and q values? What can you conclude about the present population?

a. It appears that there are three genotypes in the population caused by two alleles (**109**). Since type 3 gives all types of progeny it must be the heterozygote.

Let:

$$A^F = \text{``fast'' allele}$$
$$A^S = \text{``slow'' allele}$$

Then:

$$\text{Type 1} = A^S A^S$$
$$\text{Type 2} = A^F A^F$$
$$\text{Type 3} = A^F A^S$$

The situation is slightly different from the gel in **108** because the heterozygote shows an intermediate band. This is explained by the protein being a dimer: in a heterozygote the two different alleles direct the synthesis of different monomers which combine in all possible ways at random, as follows:

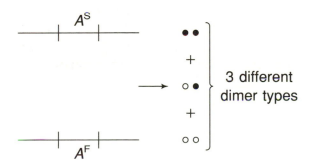

b. Principle **109** shows how to calculate p and q if all the genotype classes are distinguishable. Here there is a kind of codominance (**6**) so all three types *are* distinguishable. (This calculation makes no assumptions about Hardy-Weinberg equilibrium). We see that, letting p = frequency of A^F, and q of A^S,

$$p = (408 + \tfrac{1}{2}(384)) \div 1000 = 0.6$$
$$q = (208 + \tfrac{1}{2}(384)) \div 1000 = 0.4$$

c. If the population was in Hardy-Weinberg equilibrium, the genotypic proportions (**110**) should be:

$$A^F A^F = p^2 = 0.36$$
$$A^F A^S = 2pq = 0.48$$
$$A^S A^S = q^2 = 0.16$$

These are obviously not the frequencies in the population, so it is *not* in equilibrium.

d. This section gives us a possible reason why the population structure is not as simple as expected. Principle **117** tells us that with inbreeding the proportion of heterozygotes is:

$$2pq(1 - F) = 2 \times 0.6 \times 0.4 \times 0.8 = 0.384$$

This is the precise frequency seen in the present population, so inbreeding could be the cause of the deficiency of heterozygotes, but does not provide proof that inbreeding is the cause.

21. In a certain population of annual plants the genotypic frequencies are:

    AA   10 percent

    Aa   60 percent

    aa   30 percent

After three generations of complete selfing, what
percentage will be *AA*, *Aa*, and *aa*?

SOLUTION ——————————————————————————

Principle **117** shows that selfing reduces the proportion
of *Aa* by half each generation.

0.60 *Aa* → selfed → 0.30 *Aa* → 0.15 *Aa* → 0.075 *Aa*

The selfing process contributes equally to *AA* and *aa*, so
$0.600 - 0.075 = 0.525$ will be divided equally as 0.2625
each of *AA* and *aa*. These must be added to the original
AA and *aa* proportions, so eventually

$$AA = 0.1000 + 0.2625 = 0.3625 = 36.25\%$$
$$Aa = 0.0750 = 7.50\%$$
$$aa = 0.3000 + 0.2625 = 0.5625 = 56.25\%$$

$$\overline{}$$
$$1.0000$$

22. In *Neurospora,* the nuclei of heterokaryons never fuse. A heterokaryon is constructed between a haploid culture of genotype *nic3* and another of *ad3*. This heterokaryon is then crossed as female parent to a haploid wild-type strain acting as male. What ascus types are expected? (*nic3* and *ad3* are on different chromosomes.)

SOLUTION ────────────────────────────────────

Since the *ad3* and *nic3* genes are in separate nuclei that never fuse we expect two different meioses

$$♀ \; ad3 \times +♂$$

and

$$♀ \; nic3 \times +♂$$

The asci will be

ad3
ad3
+
+

and

nic3
nic3
+
+

(plus various second division segregation patterns **32**) reflecting Mendel's first law **8** . There can be *no* asci with both *ad3* and *nic3* segregations such as

ad3	+
ad3	+
+	*nic3*
+	*nic3*

298

23. In a *Neurospora* cross

$$A\ ad3\ nic2 \times a + +$$

The markers are linked in the order shown. From this cross most asci were as expected, with the exception of one ascus of the following type.

A	ad3	nic2
A	ad3	nic2
A	ad3	nic2
A	ad3	nic2
a	+	+
a	ad3	+
a	+	+
a	+	+

Further tests showed that all the *ad3* alleles in the five ascospores were identical at the molecular level. Propose an origin of this ascus.

SOLUTION

This is, by definition, gene conversion occurring at meiosis **34** . A random mutation is unlikely because the exceptional *ad3* allele is identical to its companions in the ascus. Accidental contamination is unlikely because the other makers are segregating properly. The gene conversion probably occurred by correction of heteroduplex DNA to mutant in a noncrossover heteroduplex structure **33** .

24. The following pedigree shows the inheritance of a muscle degenerative disease in humans. Deduce the probable mode of inheritance.

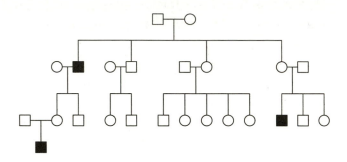

SOLUTION ──────────────────────────────────

Only men are affected, and the condition is transmitted by women who are not affected **14** . Therefore it is almost certainly an X-linked recessive condition, stemming from a gene in the original woman.

25. In dogs, the black pigment melanin is synthesized via several intermediate compounds that themselves are not pigments. A dog owner had his purebred white poodle bitch bred to an allegedly pure-breeding white stud male, expecting white puppies. However, all six puppies in the litter were black! Knowing that black is a dominant phenotype, the owner sued the owner of the stud male, accusing him of substituting another dog. The judge awarded in favour of the stud owner.

If you had been called to court for genetic advice, what advice would you have given?

SOLUTION

The judge was wrong. Geneticists know that pigments are generally synthesized through a variety of colorless intermediates (**37**). A block at any step will produce an albino. In this case the two poodles were obviously mutant for different steps, and in the F_1, complementation **53** occurred to give black pups.

$$\begin{array}{ccc} \text{Bitch } AAbb & \times & aaBB \text{ dog} \\ \text{(white)} & & \text{(white)} \\ & F_1 \; AaBb & \\ & \text{(black)} & \end{array}$$

Possible pathway:

$$\text{white}_1 \;\xrightarrow{\;\;|\;\;}\; \text{white}_2 \;\xrightarrow{\;\;|\;\;}\; \text{black}$$
$$\begin{array}{ccc} \text{gene} & & \text{gene} \\ A & & B \end{array}$$

26. Consider the following pedigree for a rare eye disease in humans, and deduce the probable mode of inheritance.

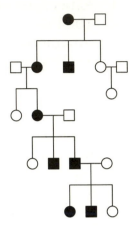

SOLUTION

The condition appears every generation (**14**) so it is most likely dominant. There is male-to-male transmission, so it cannot be X-linked; therefore, the conditions must be autosomal dominant.

27. In yeast a cross is made

$$+ \ ura3 \ + \ \times \ pro1 \ + \ leu2$$

where genes are linked in the order shown on chromosome 5 and are all auxotrophic markers. If nondisjunction occurs at the first meiotic division, what will be the composition of the resulting tetrad? (Assume that all products of meiosis develop into ascospores, and that only aneuploid ascospores with one unrepresented chromosome do not germinate.)

SOLUTION

First division nondisjunction (**65**) takes both homologues to the same pole at the first division. At the second division chromatids of both chromosomes pass into the daughter cells.

$$\underline{\underline{+ \ ura \ +}} \quad \underline{\underline{pro \ + \ leu}} \quad \nearrow \begin{bmatrix} + \ ura \ + \\ pro \ + \ leu \end{bmatrix} n + 1$$

$$+ \ ura \ + \quad pro \ + \ leu \quad \searrow \begin{bmatrix} + \ ura \ + \\ pro \ + \ leu \end{bmatrix} n + 1$$

Complementation (**53**) results in protrophy, so we expect:

2 protoprophic viable $n + 1$ spores

2 inviable $n - 1$ spores (not shown)

28. Tobacco plant cells are transformed using *Agrobacterium* T-DNA, into which has been inserted a gene for herbicide resistance. If a tobacco plant is regenerated from such a transformed cell,

   a. Draw a chromosome diagram to show what has happened to the vector
   b. If this plant is selfed what proportion of progeny will be herbicide resistant?

SOLUTION

Principle **29** shows that transformation using the T_i plasmid involves insertion of the T-DNA into a plant chromosome. Hence,

a. The transformed plant can be represented by

Inserted chromosome (I)

T DNA

Normal chromosome (N)

b. When selfed we expect (**11**)

$$\left. \begin{array}{l} \frac{1}{4} \text{ II} \\ \frac{1}{2} \text{ IN} \end{array} \right\} \text{herbicide resistant} \left(\frac{3}{4}\right)$$

$\frac{1}{4}$ NN

Herbicide resistance is predicted to be dominant because there is no equivalent DNA in the other homologue.

29. In *Drosophila*, *h* stands for hairy body and *se* stands for sepia eyes. A cross is made *se se ++* × *++ hh* and a F_1 female is crossed to a male which is *se se hh*. The offspring are as follows:

$\frac{1}{2}$ *se se +h*

$\frac{1}{2}$ *+se hh*

Propose an explanation for this result.

SOLUTION

Since both input and output genotypes are known for the F_1, it is easy to calculate recombinant frequency **22** . The two input gametes must have been *se +* and *+ h*. The two output gametic types must also have been *se +* and *+ h* (because we can subtract the *se h* contribution of the tester parent). Therefore, no recombination has occurred, and we can conclude that the loci are very closely linked. The F_1 can be represented as follows:

$$\frac{se\ +}{+\ h}$$

30. The $n + 1$ female gametophytes (embryo sacs) produced by trisomic plants are usually more viable than the $n + 1$ male gametophytes (pollen grains). If 50 percent of the functional embryo sacs of a selfed trisomic plant are $n + 1$, but only 10 percent of the functional pollen grains are $n + 1$, what percentage of the offspring will be:

a. Tetrasomic?
b. Trisomic?
c. Diploid?
d. Nullisomic?

SOLUTION

In aneuploids **65** , **67** unpaired chromosomes can pass to either pole, hence from a trisomic $(2n + 1)$, $n + 1$ gametes can form. In the present example these are differentially viable in the female and male gametophytes (which produce the gametes). Hence, fertilization can be represented as follows:

♀ \ ♂	.9 (n)	.1 ($n + 1$)
.5 (n)	.45 ($2n$)	.05 ($2n + 1$)
.5 ($n + 1$)	.45 ($2n + 1$)	.05 ($2n + 2$)

Answers therefore are:

a. 0.05
b. 0.45 + 0.05 = 0.50
c. 0.45
d. 0.

31. In corn, assume that the chemical composition of the seed can be either starchy, sugary, or waxy. Various plants were intercrossed with the following results:

cross	parents	progeny ratio		
		starchy	sugary	waxy
1.	sugary × starchy	1	0	0
2.	waxy × sugary	0	$\frac{1}{2}$	$\frac{1}{2}$
3.	waxy × waxy	$\frac{1}{4}$	0	$\frac{3}{4}$
4.	starchy × starchy	$\frac{3}{4}$	$\frac{1}{4}$	0
5.	waxy × starchy	$\frac{1}{4}$	$\frac{1}{4}$	$\frac{1}{2}$

a. Deduce the inheritance of these phenotypes. State your hypothesis and draw out each cross with genotypes.

b. Could the waxy parent in cross 2 have been the same plant as either of the waxy plants used in cross 3?

SOLUTION

a. Each cross gives a single-gene Mendelian ratio (**11**), so sugary, waxy, and starchy are probably determined by alleles (**2**). Cross 3 shows that waxy is dominant to starchy, cross 4 shows that starchy is dominant to sugary, so the allelic dominance series must be:

$$\text{waxy} \longrightarrow \text{starchy} \longrightarrow \text{sugary}$$
$$(a^w) \qquad\qquad (a^t) \qquad\qquad (a^u)$$

1. $a^u a^u \times a^u a^u \longrightarrow$ all $a^u a^u$

2. $a^w a^u \times a^u a^u \longrightarrow \frac{1}{2} a^u a^u, \frac{1}{2} a^w a^u$

3. $a^w a^t \times a^w a^t \longrightarrow \frac{1}{4} a^t a^t, \frac{3}{4} a^w -$

4. $a^t a^u \times a^t a^u \longrightarrow \frac{3}{4} a^t -, \frac{1}{4} a^u a^u$

5. $a^w a^u \times a^t a^u \longrightarrow \frac{1}{4} a^t a^u, \frac{1}{4} a^u a^u, \frac{1}{2} a^w -$

b. No, as shown in part a.

32. When a woman's karotype is tested it is found that most of her cells are XX as expected, but some are XXX and XO. What is the most likely cause of this anomaly?

SOLUTION

Principle **65** showed that aneuploids are caused by nondisjunction, and as we have a mixture of rare aneuploid and common normal tissues in this case (a mosaic; see **46**) nondisjunction must have occurred in a mitotic division at a relatively late state of development.

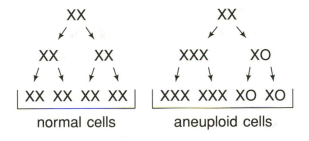

33. In humans, assume that the frequency of X-linked recessive red/green colorblindness in men is 1 in 10. What proportion of matings are capable of producing a red/green colorblind child?

SOLUTION

Principle **119** reveals the formulas that we need. We see from the men that q must be $1/10$ $(= 0.10)$, so $p = 1 - 0.1 = 0.9$.

Now, in women the genotype frequencies must be p^2, $2pq$ and q^2, and marriages are

		CC	Cc	cc
♂ \ ♀		.81	.18	.01
C	0.9		×	×
c	0.1		×	×

In the table, × represents marriages in which colorblind offspring are expected, with frequencies

$$0.9 \times 0.18 = 0.162$$
$$0.9 \times 0.01 = 0.009$$
$$0.1 \times 0.18 = 0.018$$
$$0.1 \times 0.01 = 0.001$$

Total $= 0.190$ or 19% of marriages.

34. In corn the forward-mutation rate of normal (Wx) to the waxy (wx) allelle is higher than the reversion rate (5×10^{-5} forward, 8×10^{-6} reverse).

    a. If these alleles are both neutral at the population level, what equilibrium values of p and q would result?

    b. If both rates are doubled, what effect would there be on the equilibrium values of p and q?

SOLUTION

    a. We are told:

$$Wx \xleftarrow[v = 8.10^{-6}]{u = 5.10^{-5}} wx$$

Principle **114** shows that the expected equilibrium level for these alleles is

$$\hat{q} = \frac{u}{u + v} = \frac{5 \times 10^{-5}}{5 \times 10^{-5} + 8 \times 10^{-6}} = 0.86$$

and therefore $\hat{p} = 0.14$.

Notice that in the absence of selection, the higher forward-mutation ratios inherently result in very large frequencies of mutant alleles.

    b. If all the parts of the equation are doubled, the two's are cancelled.

$$\hat{q} = \frac{2 \, (u)}{2 \, (u + v)} = \frac{u}{u + v}$$

So the answer is: no effect.

35. What progeny genotypes are expected when *Aa Bb Cc* is testcrossed to *aa bb cc*? (Assume all genes are on separate chromosomes.)

SOLUTION

Principle **8** shows that for each heterozygous gene pair, $\frac{1}{2}$ the gametes will carry one allelle and $\frac{1}{2}$ the other allele. Principle **9** shows that for independent genes, these proportions are combined randomly. Hence, the gametes from the heterozygous individual will be in the following proportions:

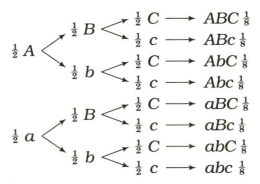

Because the tester is completely recessive, we expect these same proportions as progeny phenotypes.

36. The three pedigrees shown below all deal with the same autosomal recessive-genetic abnormality. This abnormality afflicts one out of every 160,000 people.

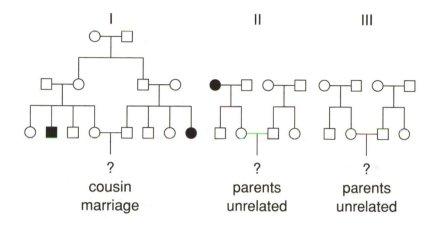

a. For *each* of these three pedigrees, estimate the probability that the first-born child of the marriage will be an abnormal daughter. (In dealing with this type of question, the convention is to apply Hardy-Weinberg rules only to families in which there is NO history of abnormality. For a family in which the abnormality occurs, the convention is to assume that those who marry into the family are homozygous for the non-mutant allele.)

b. It is found that affected people have, on the average 25 percent fewer children than do those who are normal. What is your estimate of the mutation rate?

SOLUTION ———————————————————————————————

a.   I: The mother stands a 2/3 chance of being a heterozygote as her parents were obviously heterozygotes themselves. The same is true of the father. Therefore the probability of an affected child is $2/3 \times 2/3 \times 1/4 = 4/36 = 1/9$.

II: The mother is definitely heterozygous. To calculate the probability of the father being heterozygous, we need the Hardy-Weinberg formula **110**. $q^2 = 1/160,000$, so $1 = 1/400$ and $p = 399/400$. Heterozygotes are expected at frequency $2pq = 2 \times 399/400 \times 1/400$ which is close to $1/200$. So, the probability of an affected child $= 1 \times 1/200 \times 1/4 = 1/800$.

III: In this cross the probability will be the average for the population, or 1 in 160,000.

b.   Principle **115** shows us the necessary formula.

$$q = \sqrt{\frac{\mu}{s}}$$

Hence,

$$\mu = q^2 s = \frac{1}{160,000} \times \frac{1}{4} = 1 \text{ in } 640,000 \text{ gametes}$$

37. In the fungus *Sordaria* a linear ascus analysis is made of the cross between a wild-type strain and a doubly auxotrophic strain requiring both adenine (*ad*) and tryptophane (*tr*). The results were as follows:

1	2	3	4	5	6	7	8
ad tr	ad +	ad tr	ad tr	ad tr	ad +	ad tr	ad tr
ad tr	ad +	ad +	+ tr	+ +	+ tr	+ +	ad tr
+ +	+ tr	+ tr	ad +	ad tr	ad +	+ tr	+ tr
+ +	+ tr	+ +	+ +	+ +	+ tr	ad +	+ +
36	1	39	21	1	1	1	1

a. Ignore ascus type 8 for the time being and, using only the ascus types 1 through 7, determine the linkage relations of the genes and their respective centromere(s). Be sure to draw a map to summarize your findings.

b. Draw the meiotic event(s) needed to produce ascus types 5 and 7.

c. What does ascus type 8 demonstrate? Explain very briefly.

SOLUTION ————————————————————————————

   a. We can use the frequency of second division segregation patterns to determine centromere distances **32** .

> For *ad*, we total 21 + 1 + 1 + 1 = 24, which is 12 m.u.
>
> For *tr*, we total 39 + 1 + 1 + 1 = 42, which is 21 m.u.

Inspection of the data shows that generally when a crossover occurs between one locus and the centromere (a second-division segregation) the other locus does not show a second-division segregation. Therefore it is likely the loci are in opposite arms.

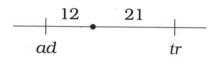

                       12   21
                 ad   *tr*

b. 5. *ad*   *tr*   7. *ad*   *tr*

          +   +   +   +

   c. Gene conversion of + to *tr* **34** .

38. In a tetraploid plant the A and B loci are centromere-linked and on separate chromosomes. A cross is made *AAAA bbbb* × *aaaa BBBB* and an F$_1$ is obtained which is then selfed to give an F$_2$. If only one dominant allele is needed to give the dominant phenotype, what phenotypes are expected in the F$_2$ and in what proportions?

SOLUTION

The F$_1$ must be *AAaa BBbb*. If the loci are on separate chromosomes, the tetraploid ratios from selfing (see **67**) can be combined randomly **11** . Hence,

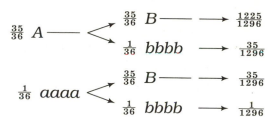

So a phenotypic ratio of 1225 : 35 : 35 : 1 is produced.

39. In a large natural population of *Mimulus guttatus*, a leaf was sampled from a large number of plants. The leaves were crushed and run on an electrophoretic gel. The gel was then stained for a specific enzyme X. Six different banding patterns were observed as shown below in the frequencies indicated.

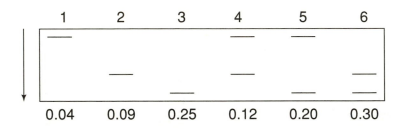

a. Assuming these patterns are produced by a single locus, propose a genetic explanation for the six types.
b. How would you test your idea?
c. What are the allele frequencies in this population?
d. Is the population in Hardy-Weinberg equilibrium?

a. This is an example of enzyme polymorphism **108** . Each plant can have either one or two bands, and there are three band positions, so it is likely that there are three alleles. We can label these A^S (slow), A^F (fast) and A^I (intermediate). Then $1 = A^S A^S$, $2 = A^I A^I$, $3 = A^F A^F$, $4 = A^S A^I$, $5 = A^S A^F$, and $6 = A^I A^F$.

b. Self the supposed heterozygotes and look for $1 : 2 : 1$ Mendelian ratios **11** in the progeny.

c.
$$A^S = 0.04 + \tfrac{1}{2}(0.12) + \tfrac{1}{2}(0.20) = 0.2 = p$$
$$A^I = 0.09 + \tfrac{1}{2}(0.12) + \tfrac{1}{2}(0.30) = 0.3 = q$$
$$A^F = 0.25 + \tfrac{1}{2}(0.20) + \tfrac{1}{2}(0.30) = 0.5 = r$$

d. Yes, the observed frequencies fit

$$(p + q + r)^2 = p^2 + q^2 + r^2 + 2pq + 2pr + 2qr.$$

40. In the synthesis of the amino acid tryptophan in *E. coli* a number of different mutations, 1, 2, 3, and 4 affect one step in this process by interfering with the production of one particular enzyme. These mutant stocks may be mated together through transduction to produce wild-type recombinants (+) or not (0) as shown in the grid below.

	1	2	3	4
1	0	0	0	+
2	0	0	+	0
3	0	+	0	+
4	+	0	+	0

a. Considering that one or all of these mutations may be deletions, draw a possible deletion map of this area.

b. If a point mutation produced wild-type recombinants with 3 and 4, but not 1 or 2, at which position on your map would it most likely be located? (Mark it with an arrow.)

SOLUTION

a. Mutants 1 and 2 fail to produce recombinants with two other mutants so are most likely deletions. Mutants 3 and 4 could be point mutations but we will assume they are deletions. Principle **60** shows how to construct a deletion map. In the following diagram the lines show the extent of the deleted regions, and overlapping lines represent failure to produce wild-type recombinants.

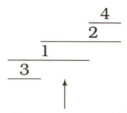

b. The arrow marks the site of the point mutation, that is, in a region spanned by deletions 1 and 2.

41. The petals of *Collinsia grandiflora* are normally blue. The blue pigment is synthesized in the cells in a biochemical pathway involving a yellow and a pink chemical intermediate as follows:

colourless precursors $\xrightarrow{}$ yellow pigment $\xrightarrow{1}$ pink pigment $\xrightarrow{2}$ blue pigment

Each step is controlled by one specific enzyme each coded by a separate gene.

A pure-breeding plant, homozygous for a gene that causes an inactive enzyme 1, is crossed to a pure-breeding plant homozygous for a gene that causes an inactive enzyme 2.

a. What color do you expect these plants to be?
b. What colour would the F_1 be?
c. If the F_1 is selfed, what ratio of phenotypes (colours) do you predict in the F_2?

Define any gene symbols you use.

SOLUTION

a. Let gene-controlling step 1 be *A*, and step 2 *B* **37** . Then first plant is *aa BB*, and is expected to be yellow **19** . The second plant is *AA bb* and is expected to be pink.

b. The F_1 will be *Aa Bb* and will be blue because of complementation of *A* and *B* **53** . In other words, both active genes are present in the F_1 dihybrid.

c. The standard Mendelian expectation at the phenotypic level **11** is

$\frac{9}{16}$ *A– B–*   blue

$\frac{3}{16}$ *A– bb–*   pink

$\frac{3}{16}$ *aa B–*   yellow $\left.\vphantom{\begin{matrix}a\\a\end{matrix}}\right\}$ $\frac{4}{16}$

$\frac{1}{16}$ *aa bb*   yellow

This is a modified Mendelian ratio **38** of $9 : 3 : 4$.

42. In mice the recessive allele a produces an albino phenotype and at another gene pair the recessive allele c produces curly hair. A pure-breeding pigmented, curly-haired mouse is crossed to a pure-breeding albino with straight hair. The F_1 is testcrossed and the progeny of the testcross examined.

- **a.** Give the genotypes of the two parental strains, the F_1, and the tester.
- **b.** Assuming the gene pairs to be unlinked, show what *genotypes* and *phenotypes* are expected in the testcross progeny, and give their *frequencies*.
- **c.** Repeat part b, assuming the gene pairs to be linked 28 cM apart.

SOLUTION

a. $aa\ ++ \times ++\ cc \rightarrow F_1\ +a\ +c$. Tester $aa\ cc$.

b. Testcross results, based on Mendel's first and second laws **9 10** .

$$+a\ +c \times aa\ cc \longrightarrow \tfrac{1}{4}\ +a\ +c$$
$$\tfrac{1}{4}\ +a\ cc$$
$$\tfrac{1}{4}\ aa\ +c$$
$$\tfrac{1}{4}\ aa\ cc$$

c.

$$\frac{+\quad c}{a\quad +} \times \longrightarrow$$

Predicted gametes based on **26**

+	c	36%
+	+	14%
a	c	14%
a	+	36%

$\left.\begin{array}{c} \\ \\ \end{array}\right\}$ 28%

When fertilized by the tester gametes $a\ c$, the same ratio is expected in the progeny.

43. *Arabidopsis thaliana* is a very small cruciferous plant that is becoming popular for studies in molecular genetics. It is relatively straightforward to obtain new interesting recessive phenotypes, which must then be mapped to other existing genes. Let's assume we have a plant homozygous for a new allele, *aa*, and we wish to determine if it is linked to the established locus, *b*. We could cross *aa* ++ × ++ *bb*, obtain dihybrids, and testcross them to *aa bb*, but because of the small flower size, controlled testcrosses are laborious and it is not always possible to prevent some selfing. A faster alternative method is to allow the dihybrids to self, select F_2 individuals of genotype *aa*, then determine what proportion of these are also *bb*.

a. What proportion of the F_2 individuals would be expected to be *bb* if the loci are unlinked?

b. Would you expect more or less if the loci are linked? Explain your answers.

SOLUTION

a. Mendel's second law **9** tells us that if the loci are independant (unlinked) then $\frac{1}{4}$ of the *aa* plants will be also *bb* [the $\frac{1}{16}$ of the $9:3:3:1$ ratio **11**].

b. If there is linkage the situation is represented as

$$\frac{a \quad +}{+ \quad b} \qquad \text{(see } \textbf{27} \text{)}$$

and obviously the *ab* combination needed to generate *aa bb* individuals in the F_2 will be much rarer, requiring crossing-over **24** to produce them.

44. Two pure-breeding lines of *Nicotiana* (tobacco) are intercrossed to form an F_1. The variance of corolla length in the F_1 is 5. The F_1 is selfed to give an F_2 whose corolla length shows a continuous range of lengths whose variance is 45. Use these data to estimate the broad-sense heritability of corolla length in these plants.

SOLUTION ─────────────────────────────────

The F_1 must be genetically uniform as it is derived from two homozygous lines, whatever their genotypes; therefore, the variance in the F_1 must be all environmental. In the F_2 the heterozygous loci all segregate and the variance has both an environmental and a genetic component.

To calculate H^2 (**104**) we need S_g^2 and S_e^2. So $S_e^2 = 5$, and S_g^2 must be $45 - 5 = 40$. Now

$$H^2 = \frac{S_g^2}{S_g^2 + S_e^2} = \frac{40}{45} = 88.88\%$$

45. *In Drosophila* the recessive allele *s* produces sable body colour (normally brown), and the recessive allele *g* produces garnet eyes (normally red). A cross of a wild-type female to a sable garnet male produced an F_1 of wild-type phenotype. When F_1 siblings were intercrossed the resulting F_2 was composed of wild-type females, and the males were one-half wild-type and the other half were of sable garnet phenotype. Explain these results.

SOLUTION

The F_1 is expected from standard Mendelian genetics, but the F_2 shows a deviation from Mendelian expectations: one $\frac{1}{4}$ of the F_2 is sable and garnet, but this quarter is all male. This is clear evidence of sex linkage of both genes **13** . Since there are no F_2 individuals that were sable or garnet alone, the two loci must be very closely linked so that crossing-over **24** is very rare.

$$
\begin{array}{ccc}
\dfrac{+}{+} \dfrac{+}{+} & \times & \dfrac{g}{s}\Big|_{Y}
\end{array}
$$

$$\downarrow$$

$$
F_1 \qquad \dfrac{+}{+} \dfrac{g}{s} \qquad \dfrac{+}{+}\Big|_{Y}
$$

$$
F_2 \qquad \underset{\male}{\dfrac{+}{+} \dfrac{+}{+}} \qquad \underset{\female}{\dfrac{+}{+} \dfrac{g}{s}} \qquad \underset{\male}{\dfrac{+}{+}\Big|_{Y}} \qquad \underset{\male}{\dfrac{g}{s}\Big|_{Y}}
$$

46. If heterozygous (*Aa*) beans are planted and allowed to self, what proportion of progeny will be heterozygous? If these individuals are again allowed to self, what proportion of the total progeny will be heterozygous? Repeat for two more generations. If possible devise a formula to show the proportion of heterozygotes after *n* generations of selfing.

SOLUTION ────────────────────────────────────

Mendelian inheritance **11** and the principle of inbreeding **117** show that when *Aa* selfs, half the progeny are *Aa*. At each generation of selfing, the proportion of heterozygotes is reduced by another half. Therefore the formula will be

$$\text{Proportion of } Aa = \tfrac{1}{2}^n$$

where *n* = the number of generations of selfing.

47. In humans, rare individuals are found that have the following sex chromosome abnormalities: XO, XXX, XYY, XXY, XXXX, XXXY, and XXXXY.

    a. What sex would these individuals be?
    b. How many Barr bodies would the cells have in each case?

SOLUTION

    a. Principle **12** tells us the sexes would be female, female, male, male, female, male, male.
    b. The purpose of X-inactivation (**61**) is to reduce the number of active X chromosomes to one per cell, therefore the number of Barr bodies predicted is 0, 2, 0, 1, 3, 2, and 3, respectively.

48. In repeated crosses between Manx cats (tailless), the litters show $\frac{2}{3}$ Manx and $\frac{1}{3}$ normal kittens. Propose an explanation for this ratio and say how to test it.

SOLUTION

Principle **38** tells us that modified Mendelian ratios provide clues about gene action and interaction.

The $2:1$ ratio is probably a vestige of a $1:2:1$ ratio (**11**), therefore it is most likely due to the death of one of the homozygous genotypes. If we let M stand for Manx, and m for normal, it is reasonable to propose that MM is a lethal genotype. Hence

$\frac{1}{4}$ MM lethal

$\frac{1}{2}$ Mm Manx

$\frac{1}{4}$ mm normal

This hypothesis can be tested by:

1. Crossing Manx by normal: a $1:1$ ratio is predicted.

2. Looking for aborted fetuses in utero in Manx $\times$ Manx crosses.

49. A new active transposon (Tn) was introduced into a haploid yeast culture by transformation. From this culture a methionine auxotrophic mutation (*met*) was isolated that was presumed to have arisen by Tn insertion into the *met*$^+$ gene. The *met* strain was crossed to wild type and the DNAs of 20 *met* progeny were probed using a Tn-specific probe. Eleven *met* progeny showed homology to the probe, and nine did not. Was the *met* mutation caused by Tn insertion?

SOLUTION —————————————————————————

Principle **95** shows that if a Tn insertion occurs, then the mutation caused by disrupting the gene's sequence, and the Tn itself, should cosegregate in 100 percent of cases. In this experiment the Tn is obviously segregating independently of *met* (**9**) and therefore the appearance of the *met* mutation in the transformed culture was coincidental.

50. A haploid yeast culture was resistant to chloram-phenicol, erythromycin, and oligomycin, caused by mutant alleles of the mitochondrial DNA: cap^R, ery^R, oli^R. A petite colony arose spontaneously in this culture, and tests revealed that this petite no longer carried the determinants for chloramphenicol and erythromycin resistance but still carried the determinant of oligomycin resistance. How can this be explained?

SOLUTION ───────────────────────────────────

Petite mutations arise from deletions of the mtDNA (**103**), so it appears from principle **59** that the deletion in this example deleted cap^R and ery^R, but left oli^R.

51. A sample of normal double-stranded DNA was isolated and found to contain 18 percent guanine.

  a. What is the expected proportion of the other three bases?

  b. In this DNA what is the expected average length of fragments derived by cutting with the restriction enzyme EcoRI

target sequence 5'GAATTC
                                    CTTAAG5'

SOLUTION ───────────────────────────────────

a. Since principle **17** shows that G = C and A = T, we can deduce that

$$G = C = 0.18$$

so

$$A + T = 1 - (2 \times 0.18) = 0.64$$

and

$$A = T = 0.32$$

Hence,

$$G = 0.18$$
$$C = 0.18$$
$$A = 0.32$$
$$T = 0.32$$

b. Principle **82** shows that restriction enzymes are highly specific in the sites at which they cut, so EcoRI will cut only at the sequence 5′GAATTC3′ in either strand. The expected frequency of this sequence anywhere in the DNA is:

| G | A | A | T | T | C |

$$0.18 \times 0.32 \times 0.32 \times 0.32 \times 0.32 \times 0.18 =$$
$$3.4 \times 10^{-4}.$$

In other words, an average of 3.4 cutting sites in 10^4 nucleotides or one site in $\dfrac{10^4}{3.4} = 2{,}941$ nucleotides.

Hence, the average sized fragment will be 2.941 kb.

52. A fruitfly of genotype *Ss Hh rr* is crossed to another of genotype *Ss hh Rr*, where all genes are unlinked. In the progeny, what will be the proportion of

    a.   *SS Hh Rr*
    b.   *Ss Hh rr*
    c.   *ss hh rr*
    d.   phenotype S– H– R–

SOLUTION

For any locus, there are very few standard ratios possible in a cross **11** . Principle **9** shows that for independant genes, the ratios may be combined randomly.

a. $Ss \times Ss \rightarrow \frac{1}{4} SS$

      $Hh \times hh \rightarrow \frac{1}{2} Hh$ Hence $SS\ Hh\ Rr = \frac{1}{4} \times \frac{1}{2} \times \frac{1}{2} = \frac{1}{16}$

      $rr \times Rr \rightarrow \frac{1}{2} Rr$

b.   *Ss Hh rr* will be at a proportion $\frac{1}{2} \times \frac{1}{2} \times \frac{1}{2} = \frac{1}{8}$

c.   *ss hh rr* will be at a proportion $\frac{1}{4} \times \frac{1}{2} \times \frac{1}{2} = \frac{1}{16}$

d.   The S–H–R– phenotype is predicted at a proportion $\frac{3}{4} \times \frac{1}{2} \times \frac{1}{2} = \frac{3}{16}$

53. A hybrid is made between a diploid plant ($2n = 10$) which is homozygous recessive for five genes (*a, b, c, d, e*), each of which is located on a different chromosome and an autotetraploid of the same species which is homozygous dominant for the genes (*A, B, C, D, E*). Gametes are produced with varying numbers of chromosomes, ranging from haploid to diploid; only the gametes that are *n* or *2n* are functional.

    a. What is the probability of the hybrid plant producing a functional gamete?

    b. What is the probability of one of the functional gametes being *abcde*?

    c. What is the probability of selfing the hybrid and obtaining progeny that are tetraploid?

SOLUTION

    a. The hybrid must be triploid $3n$ formed from the $2n$ gamete of a tetraploid and the *n* gamete of a diploid **66**. Each chromosome group of 3 will segregate 2 $\longleftrightarrow$ 1 at meiosis. The probability of a diploid gamete is therefore $(\frac{1}{2})^5$, and the probability of a haploid gamete is also $(\frac{1}{2})^5$. Together, these total $(\frac{1}{2})^4$.

    b. For any chromosome group of 3, for example, $A/A/a$, there will be three equally likely segregations
1 2 3
12 $\longleftrightarrow$ 3, 13 $\longleftrightarrow$ 2, and 23 $\longleftrightarrow$ 1. This results in a frequency of *a* gametes of $\frac{1}{6}$. Considering all the chromosomes, the frequency of *a b c d e* will be $(\frac{1}{6})^5$ **9**.

    c. Tetraploid progeny can only be obtaining by $2n$ eggs fusing with $2n$ sperm, and the probability of this is $(\frac{1}{2})^5 \times (\frac{1}{2})^5 = (\frac{1}{2})^{10}$.

54. In *Aspergillus*, a diploid of constitution

$$w/+, \ a/+, \ b/+, \ c/+, \ d/+, \ e/+, \ f/+$$

was subjected to haploidization and white (w) haploids were selected. These white haploids were then scored for the other markers and the following genotypes were found in approximately equal frequencies:

$$w + b\ c + e\ f$$
$$w\ a\ b\ c + e\ f$$
$$w + b\ c\ d\ e\ f$$
$$w\ a\ b\ c\ d\ e\ f$$
$$w + + c + + +$$
$$w\ a + c + + +$$
$$w + + c\ d + +$$
$$w\ a + c\ d + +$$

Arrange the seven genes into chromosomes.

SOLUTION

Haploidization is described in principle **35** . Chromosomes are lost randomly, so if all genes were unlinked, a total of $(2)^6 = 64$ genotypes would be expected. The fact that we have eight genotypes suggests a total of three chromosomes in addition to that bearing w. Since all bear c, then c must be linked to w, the selected marker. Apparently b, e, and f must be linked because only two genotypes result for those loci. The loci a and d appear to be lost independently. In conclusion, the chromosomal groups are

$$\begin{array}{c} b \\ w\ e \\ \underline{c\ f}\ \ \underline{a}\ \ \underline{d} \end{array}$$

55. Assume that in broccoli, cytoplasmic male sterility (*cms*) is correlated with a specific mitochondrial DNA restriction-enzyme fragment pattern different from fertile individuals. Because *cms* is agriculturally useful, you want to transfer the *cms* trait to an established line L with a favourable nuclear genotype, that is presently male fertile.

    a. One way to transfer cytoplasms is by repeated backcrosses. Say how this would be done in the present case and discuss some of the problems that might be associated with this technique.

    b. Another technique is to use cell fusion. Assume you have a way of removing cell walls, fusing cells, etc., how might you isolate a cms derivative of the favourable line? (Hint: it is possible to effectively enucleate cells using ionizing radiation)

SOLUTION

    a. Cross *cms* females $\times$ L males, then cross F_1 as female $\times$ L male, and repeat as many times as practically possible. Maternal inheritance of *cms* **99** ensures that an individual will be produced with *cms* cytoplasm containing a predominantly L nuclear genotype. The problems with this method are that it is tedious and time-consuming.

    b. Fuse enucleated *cms* cells with L cells and allow the cells to divide to see if cytoplasmic sorting-out occurs **100** . After several cell cycles regenerate plants from the tissue **68** and some should have *cms* cytoplasm with an L nucleus.

56. A human DNA probe G8 was found to pick up a restriction-fragment-length polymorphism (RFLP). Pedigree studies showed that this RFLP showed linkage to Huntington's disease (HD).

In a separate line of experiments it was shown that patients with Wolf-Hirshhorn syndrome (WHS) (a form of mental retardation) all have variously sized deletions of the tip of the short arm of one of their chromosomes 4.

It had been suspected for various other reasons that the gene for HD is located on the short arm of chromosome 4. How would you use the above information to test that idea?

SOLUTION

Deletions **59** result in loss of DNA that normally resides in that region so the individuals with WHS should all be hemizygous (**13**) for at least part of the short arm of chromosome 4. If the deleted region is the one to which G8 hybridizes, then such individuals should show only one RFLP haplotype **108** , whereas normal individuals can show two.

57. In corn, a gene C is needed for purple anthocyanin production in the seed. In the absence of C function, the seed is yellow. An autonomously active transposon Tnl inserts into the middle of the C gene to give the allele C^{Tnl}. This transposon exits frequently and late in development

    a. What will be the phenotype of seeds of genotype $C^{Tnl}C^{Tnl}$?

    b. Another transposon Tn2 exists rarely and early in development; what will be the phenotype of seeds of genotype $C^{Tn2}C^{Tn2}$?

    c. What will be the phenotype of seeds of genotype $C^{Tnl}C^{Tn2}$?

Principle **58** shows that transposons can inactivate a gene, and the exit of a transposon during the development of somatic tissue such as a seed produces a revertant sector (see also **46**).

a. We expect the following

b. We expect the following

c. Because the exit of the two different transposons is independent, we expect

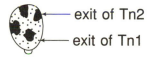

exit of Tn2

exit of Tn1

58. Wood ducks may have short legs or long legs. They also may have normal tails, pointed tails, or fan tails. A short-legged duck with a pointed tail was mated with a long-legged duck that had a normal tail. All the offspring had short legs and fan tails.

These F_1 progeny were allowed to interbreed, and the offspring showed *only* the following characteristics in the proportions noted:

Short-legged, normal-tailed   3

Short-legged, fan-tailed   6

Short-legged, pointed-tailed   3

Long-legged, normal-tailed   1

Long-legged, fan-tailed   2

Provide the simplest explanation for the unexpected ratio, and for the inheritance and expression of characteristics in these results. (Courtesy of I.E.P. Taylor)

SOLUTION ——————————————————————————————

From the F_1 phenotypes, it appears that short legs (L) are dominant **6** , and that fan tail is a case of a new phenotype associated with a heterozygote, say P^1P^2.

Possible genotypes: Parents $LL\ P^1P^1 \times ll\ P^2P^2$
$$F_1 \qquad Ll\ P^1P^2$$

Predicted F_2 **8 9** $\frac{3}{4}\ L-$ $\longrightarrow$
- $\frac{1}{4}\ P^1P^1$ short, normal   3
- $\frac{1}{2}\ P^1P^2$ short, fan   6
- $\frac{1}{4}\ P^2P^2$ short, pointed   3

$\frac{1}{4}\ ll$ $\longrightarrow$
- $\frac{1}{4}\ P^1P^1$ long, normal   1
- $\frac{1}{2}\ P^1P^2$ long, fan   2
- $\frac{1}{4}\ P^2P^2$ long, pointed   1

Obviously the last class is missing, presumably due to a lethal interaction between the long and pointed determinants.

59. In *E. coli* the gene *lacZ* enters the F⁻ cell about one minute ahead of *phoR*. In a cross Hfr ++ × F⁻ *lacZ phoR*, *phoR*⁺ exconjugants are selected and tested further. It is found that 83 percent of them are *lacZ*⁺ and the remainder are *lacZ*. Explain this result and give a map distance in recombination units.

SOLUTION

The merozygotes **23** selected in this case were

	pho^+		lac^+	
1		2		3
	pho		lac	

The pho^+ lac^+ exconjugants arose from homologous exchange **24** in region 1 and 3 (see **26**), and the pho^+ lac^- types from exchanges in 1 and 2. Since principle **26** says that map distance is proportional to recombinant frequency (RF), the RF here (17%) can be translated into 17 map units.

60. A yeast *leu2* (leucine requiring) mutant was UV-irradiated and plated on minimal medium; rare prototrophic colonies were then observed and collected. They were all crossed to a standard yeast wild type with the following results:

$$\text{Most prototrophs} \atop \times \text{ wild type} \longrightarrow \text{progeny all were} \atop \text{leucine independent}$$

$$\text{One unusual} \atop \text{prototroph} \atop \times \text{ wild type} \longrightarrow \frac{3}{4} \text{ leucine independent} \atop \frac{1}{4} \text{ leu2}$$

Explain these results. (Note: all strains haploid.)

SOLUTION

Principle **39** tells us that mutant alleles can revert so the majority of the prototrophs were $leu2^+$ revertants, which on crossing to the $leu2^+$ allele in wild type gave all wild-type progeny.

The unusual prototroph obviously still harboured the *leu2* mutation, presumably with a new suppressor mutation that we can call *su* **57** . This must have segregated independantly **9** in the cross:

$su\ leu2 \times +\ + \longrightarrow$

$\frac{1}{2}\ su \Big\langle \begin{matrix} \frac{1}{2}\ leu2 \longrightarrow \frac{1}{4}\ su\ leu2 \\ \frac{1}{2}\ + \quad\ \longrightarrow \frac{1}{4}\ su\ + \end{matrix}$

$\frac{1}{2}\ + \Big\langle \begin{matrix} \frac{1}{2}\ leu2 \longrightarrow \frac{1}{4}\ \boxed{+\ leu2} \longrightarrow \text{the only genotype} \\ \text{to require leucine} \\ \frac{1}{2}\ + \quad\ \longrightarrow \frac{1}{4}\ +\ + \end{matrix}$

345

61. In maize, p = purple leaves (+ = green), v = virus-resistant (+ = sensitive), and b = brown striped seed (+ = plain). A cross is made of a purple-leaved, virus-resistant, brown-seeded plant to a wild type, and the F_1 was testcrossed to a triple recessive. The resulting progeny were:

purple, resistant, brown	1015
green, sensitive, plain	1370
green, sensitive, brown	249
purple, resistant, plain	254
green, resistant, plain	185
purple, sensitive, brown	159
green, resistant, brown	8
purple, sensitive, plain	9
	3249

Deduce the linkage arrangement of these genes and calculate interference if appropriate.

SOLUTION ——————————————————————————

We need to calculate recombinant frequencies (**26 27**)

For

p-b: 249 + 254 + 8 + 9 = 520   ⟶ 16% = 16 m.u.
p-v: 185 + 159 + 8 + 9 = 361   ⟶ 11% = 11 m.u.
v-b: 249 + 254 + 185 + 159 = 847 ⟶ 26% = 26 m.u.

Only one order is compatible with these data:

$$\underset{11\qquad\ \ 16}{\overset{v\qquad\quad p\qquad\ \ b}{\longmapsto\!\!\!\mid\!\!\!\longmapsto}}$$

$$\text{Interference} = 1 - \left(\frac{8 + 9}{3249 \times .11 \times .16}\right)$$

$$= 1 - \left(\frac{17}{57}\right) = 1 - 0.3 = 0.7 \text{ or } 70\%$$

62. In *Drosophila* the genes for ebony body (*e*) and stubble bristles (*s*) are linked on chromosome 2. A fly of genotype + *s/e* + developed with predominantly wild-type phenotype, but had a pair of adjacent patches, one of which showed stubble bristles, and the other was ebony in colour. What is the most likely origin of this mosaic?

SOLUTION

Out of all the mechanisms of generating mosaics (**46**), mitotic crossing-over (**35**) is the most likely because the characteristic twin adjacent homozygous spots are seen. Since both linked heterozygous loci are made homozygous, the crossover must have occurred between the centromere and the loci:

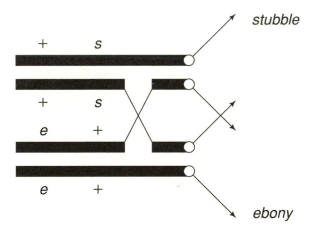

63. Why don't crosses of an allopolyploid to either of its parental species produce fertile offspring?

SOLUTION

This happens because there is always a chromosome set with no pairing partners, and a normal meiosis cannot occur. Principle **66** showed that allopolyploids are $2n_1 + 2n_2$, so if we cross this to a parent we see the following:

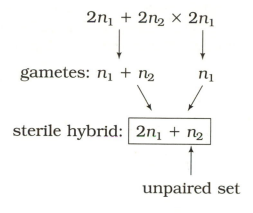

$$2n_1 + 2n_2 \times 2n_1$$

gametes: $n_1 + n_2$ $\qquad$ n_1

sterile hybrid: $\boxed{2n_1 + n_2}$

unpaired set

64. In the annual plant *Collinsia,* the leaves are normally dark green. A pure-breeding mutant line arose that had yellow leaves. When crossed to normal the F_1 was yellowish green, and the F_2 was $\frac{1}{4}$ green, $\frac{1}{4}$ yellow, and $\frac{1}{2}$ yellowish-green. Explain these results with a genetic model.

SOLUTION

The $1:2:1$ ratio is an F_2 monohybrid ratio **5 11** , so clearly one heterozygous gene pair is responsible for the results. Since the F_1 and $\frac{1}{2}$ the F_2 must be heterozygous, there is incomplete dominance **6** . The model must be:

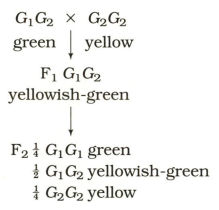

$$G_1G_2 \times G_2G_2$$
green | yellow

$F_1\ G_1G_2$
yellowish-green

$F_2\ \frac{1}{4}\ G_1G_1$ green
$\frac{1}{2}\ G_1G_2$ yellowish-green
$\frac{1}{4}\ G_2G_2$ yellow

65. In a *Neurospora* cross involving linked genes, *his nic* ×++, in addition to nonrecombinant tetrads, two types of tetrads were recovered that showed recombinants

+	+
+	nic
his	+
his	nic

and

+	nic
+	nic
his	+
his	+

(common)   (rare)

Is this result compatible with the supposition that crossing-over occurs only at the four-chromatid stage (**25**)?

SOLUTION

Yes, the second tetrad can be explained by a four-chromatid double crossover

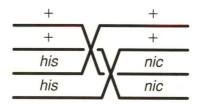

which are expected to be rare.

66. Duchenne's muscular dystrophy (DMD) is a recessive sex-linked disease. In trying to locate what region of the X might be involved, it was noticed that some DMD patients carry translocations, in which one of the breakpoints is always in band Xp21 (on the short arm of the X) and the other is in a nonspecific autosome.

    a. What does this observation suggest?

    b. One of the autosomal breakpoints is in a gene for rRNA. If a probe is available for detecting this gene for rRNA, how could the probe be used to determine the region of DNA responsible for DMD?

SOLUTION ─────────────────────────────────

    a. The DMD gene is in Xp21 and the translocation breakpoint **64** is disrupting the function of that gene.

    b. Clone the translocation genome in a library **88** and probe **89** with rRNA gene to find a clone with juxtaposed rRNA and DMD fragment. Subclone the DMD fragment and use it to recover complete DMD gene from a normal genome library.

67. Bacterial neomycin-resistance (*neo*^{r}) genes can confer neomycin resistance to mouse cells if a mouse promoter P is attached upstream. A mouse line was obtained that contained a P-*neo*^{r} construct inserted into a mouse chromosome. Mutagenesis induced a deletion in the *neo* sequence so that the cells were no longer resistant. These cells were then transformed with a circular vector carrying a P-*neo*^{r} sequence with a nucleotide pair substitution in the *neo* gene, so that this gene too was inactive. However, although most transformed cells were sensitive, resistant cells were detected at frequencies too high to be attributed to mutation. Suggest some ways in which these could have occurred and say how you would test your ideas.

SOLUTION ————————————————————————

Some possibilities are:

1. Pairing and homologous recombination (**22**) between vector gene and chromosomal gene caused insertion of entire vector flanked by a whole copy plus a double mutant (substitution plus deletion) version of the P-*neo* construct.

2. Gene conversion (**34**) replaced the deletion with normal sequence from the vector. 1 and 2 could be distinguished by restriction mapping (**82**) and/or sequencing (**93**).

3. Gene conversion replaced the mutant site on the vector. Isolate the vector from the cells and sequence the appropriate site.

68. Two children are investigated for the expression of a gene (*D*) which encodes an important enzyme for muscle development. The results of the studies of the gene and its product are given below:

Individual 1

Probe: ^{32}P Gene *D* : Autoradiograms

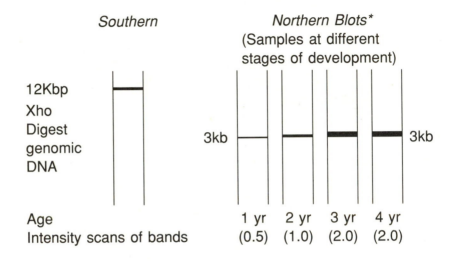

| | Southern | | Northern Blots* (Samples at different stages of development) |

		1 yr	2 yr	3 yr	4 yr
Age		1 yr	2 yr	3 yr	4 yr
Intensity scans of bands		(0.5)	(1.0)	(2.0)	(2.0)

Enzyme Samples

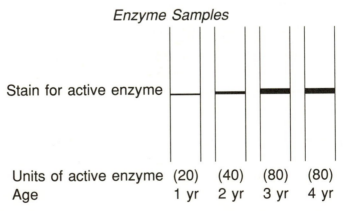

Stain for active enzyme					
Units of active enzyme		(20)	(40)	(80)	(80)
Age		1 yr	2 yr	3 yr	4 yr

*RNA electrophoresed and transferred to cellulose nitrate.

354

Individual 2

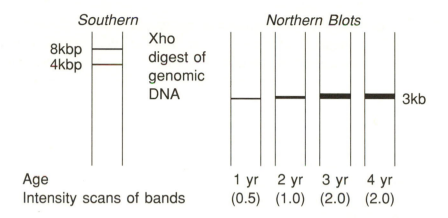

Southern

8kbp
4kbp

Xho
digest of
genomic
DNA

Northern Blots

3kb

Age
Intensity scans of bands

1 yr   2 yr   3 yr   4 yr
(0.5)   (1.0)   (2.0)   (2.0)

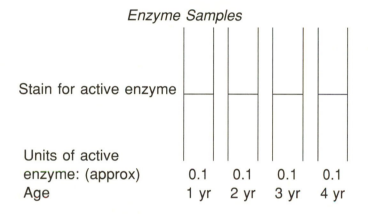

Enzyme Samples

Stain for active enzyme

Units of active
enzyme: (approx)
Age

0.1   0.1   0.1   0.1
1 yr   2 yr   3 yr   4 yr

For individual 2, the enzyme activity at each stage was very low and could only be estimated at approximately 0.1 units at age 1, 2, 3, and 4.

a. For both individuals, draw graphs representing the developmental expression of the gene. [Fully label both axes]

b. How can you explain the very low levels of active enzyme for individual 2. (Protein degradation is only one possibility!)

c. How might you explain the change in the Southern blot for individual 2 compared with individual 1.

d. If only one mutant gene has been detected in family studies of the two children, define the individual children, as either homozygous of heterozygous for the gene *D* (*d*).

SOLUTION

Graphs showing levels of RNA (transcripts) and levels of enzyme activity (units)

a. *Individual 1*

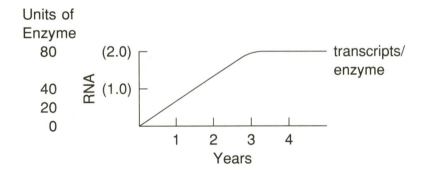

Individual 2

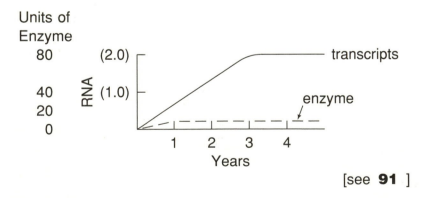

[see **91**]

b. RNA levels resemble those for the normal individual —1. Enzyme units are always very low (insignificant activity). You should therefore predict *either:* mutation in coding region (possibly active site) resulting in inactive enzyme, *or* translational mutant giving very low levels of protein.

c. The mutation results in the generation of a Xho-recognition site within gene *D*, giving two fragments.

d.   Individual 1 = homozygous *DD* [**82 83 91**]
Individual 2 = homozygous *dd* [**83**]

69. Fragments (F) of the transcribed strand of a mouse gene are given below:

F1: 3′ AGA GCC ATG TTT CCT 5′

F2: 3′ CCT TAC ACA CCA GAA 5′

F3: 3′ ACA CCA ACT CCT TTT 5′

a. Write out the RNA sequence and orientation for each region.
b. Write out the amino acid sequence and orientation (order the fragments and mark NH₂ and COOH termini)
c. Diagram the additional structural features which would be required to obtain expression of such a gene (include all stages until protein product).

SOLUTION ────────────────────────────────────

NOTE: Each sequence is continuous; the triplets provide ease of scanning.

a. F1: 5′ UCU CGG UAC AAA GGA 3′

F2: 5′ GGA AUG UGU GGU CUU 3′

F3: 5′ UGU GGU UGA GGA AAA 3′

b. There is a single AUG = Methionine start codon in F2; therefore, this represents the NH_2 terminus.

There is a single UGA = Stop codon in F3; therefore, this represents the COOH terminus. The sequence and order are:

$$\overbrace{\text{F2}}\quad\overbrace{\text{F1}}\quad\overbrace{\text{F3}}$$

NH_2 Met:Cys:Gly:Leu:Ser:Arg:Tyr:Lys:Gly:Cys:Gly COOH

[see **18 21**]

c. −30   +1

TATA ───────── initiation ─────┬──────────
                 codon   │
                 (RNA)   │   ATG
                                │   translational
                                │   start
                                │
                                │
                                │
                                │
                          ribosome
                          binding site

[see **72 79**]

359

70. DNA of virus B can exist in two different forms (type 1 and type 2). The restriction maps of these are given below. The figures represent restriction fragments of DNA electrophoresed on agarose, then stained with ethidium bromide.

Type 1

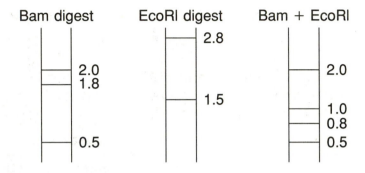

Bam digest

2.0
1.8

0.5

EcoRI digest

2.8

1.5

Bam + EcoRI

2.0

1.0
0.8
0.5

Type 2

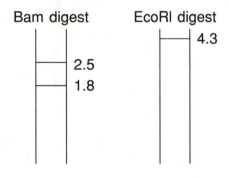

Bam digest

2.5
1.8

EcoRI digest

4.3

a. Draw the structures of types 1 and 2, marking the restriction cleavage sites.

b. DNA is isolated from a liver biopsy of a jaundiced dog that has been exposed to virus B. A sample of liver genomic DNA is investigated using a ^{32}P-labeled probe of either type 1 or type 2 viral DNA. The result is:

Genomic DNA (liver): Eco RI digest.

Southern blot, then hyridization with ^{32}P viral probe.

4.3

What does this suggest?

SOLUTION

a. Virus B, type 1

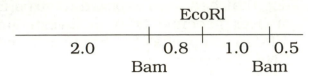

Type 2

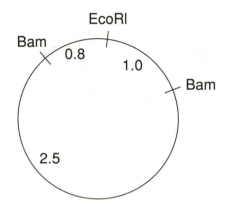

b. It suggests a circular virus DNA in liver DNA sample, which is cleaved at a single site with EcoRI. [see **3 82**]

71. A viral DNA fragment carrying a specific gene *V*, was transfected into a muscle-cell culture. Following incubation with ^{32}P-ribonucleotides, the viral-encoded RNA product was isolated at two timed intervals. The ^{32}P-viral RNA was treated as follows:

1. Hybridized to a specific cDNA previously constructed from viral gene *V* mature mRNA, then

2. The hybrid was treated with RNAase, which destroys RNA single strands.

3. The hybrid was denatured and electrophoresed on a gel that was subjected to autoradiography.

The results suggest that the pathology of the virus is time-dependent:

Results

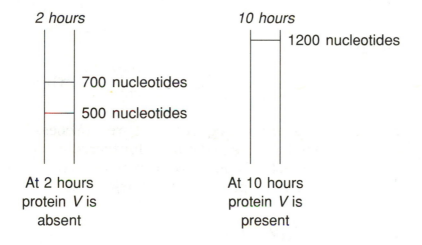

2 hours

700 nucleotides
500 nucleotides

At 2 hours
protein *V* is
absent

10 hours

1200 nucleotides

At 10 hours
protein *V* is
present

a. What is the size of the mature mRNA for gene *V*?
b. Draw a diagram of each hybrid and indicate what the above bands represent.
c. Why is protein *V* not produced until after two hours?

SOLUTION

a. 1200 nucleotides
b. *2 hours*

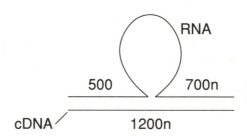

10 hours

c. Protein *V* must require mature, processed RNA (shown in hybrid at 10 hours) for translation. Unprocessed RNA (with putative intron) does not translate to give protein product (2 hours). [see **75 84 90**]

72. α-interferon is encoded by a gene which does not contain introns. The Bam H1 restriction fragment containing the complete gene can be identified on a Southern blot by hybridization to a specific interferon cDNA probe (^{32}P), which under the conditions of hybridization will only detect interferon sequences.

In order to determine the cause of unknown immune deficiencies, blood samples from patients and "normal" people are screened for the α-interferon gene and its expression.

The results are presented on the following page. Autoradiograms for Southern blots, Northern blots, and protein gels are given. The protein bands are detected by antibody which recognises α-interferon sequences. Individuals 1 and 2 are normal for immune capacity; individuals 3, 4, and 5 have immune deficiencies.

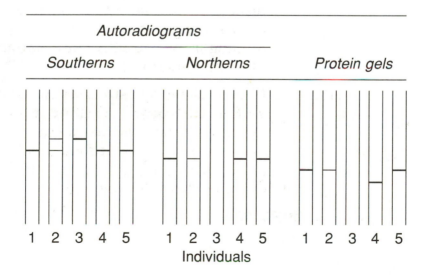

a. Which individuals are homozygous for the α-interferon gene?
b. What do you think are the causes of the immune deficiencies of individuals 3, 4, and 5? Describe the type of gene mutation involved for each of the individuals 3, 4, and 5.

SOLUTION

a. The homozygous individuals are: 1, 3, 4, 5. Individual 2 is heterozygous for the α-interferon gene, that is, for the two parental chromosomes, the DNA sequence flanking the α-interferon gene differs with regard to a Bam H1 recognition site.

b. The immune deficiencies have the following causes:

Individual 3: No specific RNA detected on Northern. Interferon genes are not transcribed; mutation is probably in the promoter.

Individual 4: RNA is the correct size. However, the protein is smaller than normal size. Mutant with a stop codon within the reading frame would produce an inactive, truncated protein.

Individual 5: RNA is the correct size, and the protein band is the correct size (and immunologically recognized as α-interferon). Lack of activity is probably due to point mutation in protein, for example, at the active site.

[see **52 70 72 83**]

73. The cDNA clone and genomic clone for a phosphatase enzyme have been isolated. From the following data the structural characteristics of the gene and its transcript can be determined.

cDNA map

The fragment of cDNA was excised from a plasmid and end-labeled with ^{32}P.

Combined digest with Hae III and Taq 1 enzymes

Electrophoretic fragments stained with ethidium bromide

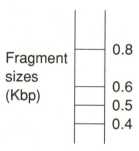

Gel from (1) exposed to X-ray film, i.e., autoradiogram pattern

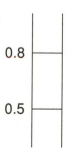

Digestion with Taq 1 enzyme alone

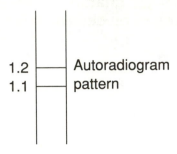

1.2 — Autoradiogram
1.1 — pattern

a. Determine the cDNA map.

Genomic DNA map

A fragment of genomic DNA was cleaved out of a λ phage clone with EcoR1 enzyme and its restriction characteristics determined. The fragment was end-labeled and digested.

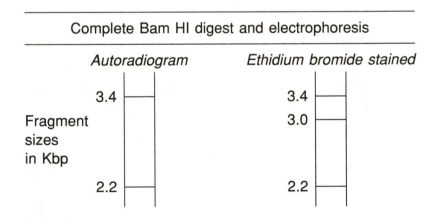

Complete Bam HI digest and electrophoresis

Autoradiogram	*Ethidium bromide stained*
3.4	3.4
	3.0
Fragment sizes in Kbp	
2.2	2.2

b. Draw the genomic map, marking the restriction sites.

c. A labeled cDNA probe hybridized to the 3.4 and 2.2 kbp genomic fragments. The 1.2 kbp Taq 1 fragment hybridized to the 3.4 kbp genomic fragment. If the phosphatase gene is single-copy which genomic fragment might the 1.1 kbp Taq 1 fragment hybridize to?

d. Which part of the gene does the 3.0 kbp genomic fragment represent?

SOLUTION

a. cDNA

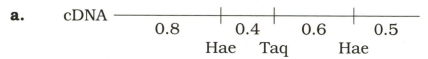

| | 0.8 | | 0.4 | | 0.6 | | 0.5 |
| | | | Hae | Taq | | Hae | |

b. Genomic

| EcoRI | | Bam | | Bam | | EcoRI |
| | 3.4 | | 3.0 | | 2.2 | |

c. The 1.1 Taq 1 fragment would hybridize to the 2.2 kbp fragment.

d. The 3.0 kbp fragment represents intron sequences; it will not hybridize to cDNA probe.

[see **82 88 90**]

74. **a.** The following analyses were carried out on the phage X4 to determine its structural characteristics. DNA isolated from a stock culture was restricted using three enzymes, and the results are given below. From these data prepare a restriction map of the phage DNA.

Agarose gels/ethidium bromide stained

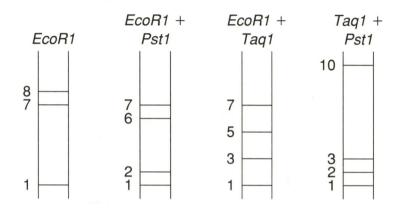

b. A digest of the stock culture phage DNA with Bam H1 alone gave the pattern shown in Figure 1. The phage was used to infect *E. coli*, passaged through several cell cycles, and DNA from the phage was then isolated. The Bam H1 restriction pattern of this DNA had changed, giving the pattern in Figure 2. What do you think has occurred?

Patterns after gel electrophoresis

Complete Bam H1 digests, gel electrophoresis, and staining with ethidium bromide

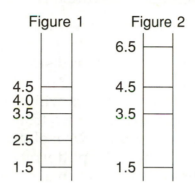

Figure 1   Figure 2

SOLUTION ————————————————————————————

a.

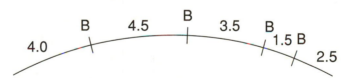

R1   R1   P   R1

7   3   3   2   1

T

[**81 82**]

b.   i.

B   4.5   B   3.5   B
4.0   1.5 B
                            2.5

ii.

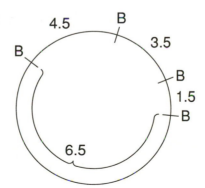

[**3**]

75. A kanamycin-resistant, ampicillin-resistant plasmid is cleaved with Pst 1 enzyme, which cleaves within the gene for ampicillin resistance. The DNA is ligated with products of Pst 1 digestion of *Drosophila* DNA, and then used to transform *E. coli*.

    a. What antibiotic would you put in the agar to ensure that a colony has the plasmid?

    b. What antibiotic-resistance phenotypes will be found on the plate?

    c. Which phenotypes will have *Drosophila* DNA?

    d. If individual colonies from the plate in part b are cultured and DNA isolated from these batches, what overall patterns might you observe following Pst I cleavage and gel electrophoresis with ethidium bromide staining?

SOLUTION

    a. Kanamycin (Kan)

    b. Kan-res and kan-res, ampicillin (amp)-res.

    c. Kan-res and amp-sensitive

    d.

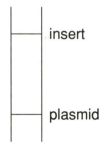

insert

plasmid

[see **81 86**]

76. Several individuals were screened for a gene (*twitch*) encoding a muscle protein. Samples of DNA + mRNA were obtained from the individuals and examined using a ^{32}P probe of the *twitch* gene. The results for individuals 1, 2, 3, and 4 are given below.

Protein products isolated from each individual were identified by specific antibody precipitation and gel electrophoresis as shown. An assessment of the resulting biological activity of this protein is indicated below each protein gel.

Southern blots of EcoR1 digests of genomic DNA.

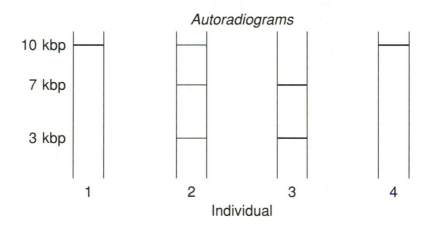

Northern blots

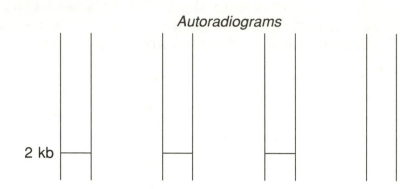

Autoradiograms

2 kb

Protein products identified by antibody against muscle protein twitch

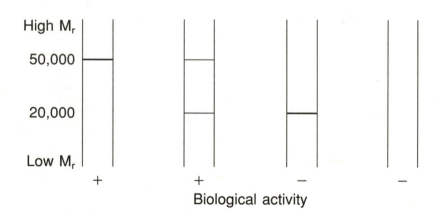

High M$_r$

50,000

20,000

Low M$_r$

$+$ $+$ $-$ $-$

Biological activity

a. Define each individual as heterozygous or homozygous for this gene.

b. Which individuals have mutant forms of the gene?

c. What types of mutants are these? Comment on the transcription and translation products.

SOLUTION ———————————————————————

   a.   Individual 1, 4, and 3 are homozygous. Individual 2 is heterozygous.

   b.   Individuals 3 and 4 have mutant forms of the gene, that is, no protein [**83 91**].

   c.   For individual 3 the RNA size and amount are correct, and the protein size and amount are correct, suggesting that there is a point mutation resulting in loss of protein activity [**52**].

       Individual 4 has no RNA. This suggests a transcription mutant, probably in the promoter region of the gene [**70**].

77. The partial sequence of a gene is given below. This DNA fragment was used as shown.

1. Ligated behind a lac promotor and the chimeric gene used to transform *E. coli* cells.

2. Ligated behind a promotor for mouse epithelial growth factor and the chimeric gene used to transform mouse epithelial cells.

DNA fragment

nucleotide number

1	9	18	27	36	45

CCTGCTGAA GTCATGCTA GCTGTAAAG AAACTATCT ACTTTTCAG

GGACGACTT CAGTACGAT CGACATTTC TTTGATAGA TGAAAAGTC

The sequence is given in groups of nine nucleotide pairs for convenience of scanning the data.

a. If stable transcripts and polypeptides were made, what length would the resulting polypeptides be? (Express in number of amino acids.)

b. If triplet 150 from the initial ATG was mutated to a translational termination codon, what length would the polypeptide products be? (Express in number of amino acids.)

c. If a mutation in the nucleotide pair at position 44 results in a change from A to C,
$$ T \quad \ G$$
what difference would be observed in the products?

```
                 900 nucleotide pairs:
                 no stop codons
GTT----------------------------  ACTTTCCAG  GATACAATC  ACTTGAAAC
CAA----------------------------  TGAAAGGTC  CTATGTTAG  TGAACTTTG
```

SOLUTION ─────────────────────────────────────

   a. The resulting polypeptides AUG to TGA would be 319 amino acids long.

   b. The polypeptide products would be 149 amino acids long.

   c. The amino acids would change to Gln $\simeq$ Pro.

[see **18 52 94**]

78. For each of the *E. coli* diploids that follow, indicate whether the strain is inducible, constitutive, or negative for β-galactosidase.

a. $i^+o^+z^-y^+/i^+o^cz^+y^+$
b. $i^+o^+z^+y^+/i^-o^cz^+y^-$
c. $i^-o^+z^-y^+/i^-o^cz^+y^+$
d. $i^+o^+z^-y^+/i^+o^cz^+y^-$
e. $i^+o^+z^-y^+/i^+o^cz^-y^-$
f. $i^+o^+z^-y^+/i^+o^+z^+y^+$

SOLUTION

a. β-galactosidase is constitutive (from $i^+o^cz^+y^+$).
b. β-galactosidase is constitutive (from $i^-o^cz^+y^-$).
c. β-galactosidase is constitutive (from $i^-o^cz^+y^+$).
d. β-galactosidase is constitutive (from $i^+o^cz^+y^-$).
e. β-galactosidase is negative (from z^-).
f. β-galactosidase is induced (from $i^+o^+z^+y^+$).

[see **71 73**]

79. Pulsed-field gel electrophoresis (PFGE) is a technique in which the electric field is applied at varying angles across the gel, enabling the separation of very long DNA molecules. In the fungus *Neurospora* ($n = 7$) PFGE was performed on carefully prepared total-cell DNA and seven clearly distinct bands were seen.

    a. What are these bands and how would you test your idea?

    b. What are some potential uses of this technique in genetics?

SOLUTION

    a. The bands are presumably the seven DNA molecules that constitute the chromosomes of this fungus. Test with a Southern blot (**91**) using probes composed of already mapped genes, for example, probes A and B might be composed of genes both known to reside on chromosome 6: both are then predicted to hybridize to the same DNA band.

    b. One obvious possibility is to determine which chromosome an unmapped cloned gene is on, a procedure that effectively by-passes traditional genetic recombination analysis.

80. A certain probe P revealed a restriction-fragment-length polymorphism (RFLP) in a natural population of *Drosophila.* The X rays revealed the following classes and frequencies:

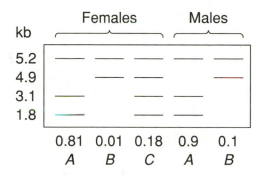

a. Interpret these results in terms of a concise hypothesis.
b. Is the population in equilibrium for the RFLP locus?
c. What types do you predict in the progeny of the following crosses
  i. *C* female × *B* male
  ii. *C* female × *A* male?
  iii. *B* female × *A* male.

a. The absence of type *C* males suggests an X-linked RFLP (**13 108**). The probe hybridizes as follows:

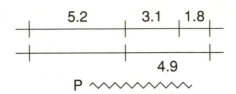

b. Yes according to the formulas for X-linked equilibria (**119**).
c.   i. Females 1/2 *C*, 1/2 *B*; males 1/2 *A*, 1/2 *B*
  ii. Females 1/2 *A*, 1/2 *C*; males 1/2 *A*, 1/2 *B*
 iii. Females all *C*; males all *B*

INDEX

Numbers refer to the principle numbers.

385